WATER RESOURCE PLANNING, DEVELOPMENT AND MANAGEMENT

GROUNDWATER

HYDROGEOCHEMISTRY, ENVIRONMENTAL IMPACTS AND MANAGEMENT PRACTICES

WATER RESOURCE PLANNING, DEVELOPMENT AND MANAGEMENT

Additional books in this series can be found on Nova's website
under the Series tab.

Additional e-books in this series can be found on Nova's website
under the e-book tab.

WATER RESOURCE PLANNING, DEVELOPMENT AND MANAGEMENT

GROUNDWATER

HYDROGEOCHEMISTRY, ENVIRONMENTAL IMPACTS AND MANAGEMENT PRACTICES

SEGUN MICHAEL ADELANA
EDITOR

nova publishers
New York

Library of Congress Cataloging-in-Publication Data

Groundwater : hydrogeochemistry, environmental impacts, and management practices / editor, Segun Michael Adelana (Victorian Department of Environment and Primary Industries, Agriculture Research Division, Australia).
pages cm
Includes index.
ISBN 978-1-63321-759-1 (hardcover)
1. Groundwater--Pollution. 2. Water chemistry. 3. Hydrogeology. I. Adelana, Segun.
TD426.G7386 2014
551.49--dc23
2014031860

Published by Nova Science Publishers, Inc. † New York

CONTENTS

PREFACE

Groundwater is important to man and the natural environment in many ways. It is one of the world's most ubiquitous, widely used resources that is unseen and misunderstood. While groundwater has a lot of prospects to supporting many human and ecosystem functions into the future, there are many problems associated with its development due to human activities or resulting from its movement through the geologic environment. Today the modern hydrogeologist is confronted with multiple reactions of water with materials in the subsurface (geogenic aquifer material) as well as surface anthropogenic sources of matter of different origin and varying concentration levels.

Many of the world's coastal aquifers are vulnerable to contamination by seawater intrusion and both primary and secondary dryland salinity have significantly degraded water quality reducing freshwater availability. The increasing demands for fresh water due to growing population coupled with the anticipated impacts of climate variability and change are expected to lead to increases in the water quality degradation. Research studies or investigations of groundwater quality degradation and the extent of monitoring and investigations specific to groundwater quality degradation are highly variable across the countries of the world. This volume was prepared to provide the relevant information along this line.

Hydrogeologists and scientists were invited to submit contributions from recent projects or research in groundwater – particularly from the quality and management perspectives of developing or managing groundwater. The result is this volume of 9 chapters, reflecting contributions with case studies from the United States, Canada, Australia, Brazil, Serbia, India, Malaysia, Netherlands and South Africa.

Nova Science Publishers and the editor would like to thank those reviewers who have spent part of their precious time to correct and comment on the various chapters, leading to improved manuscript. The following referees for the chapters are in particular appreciated: Professor José Manuel Marques (Technical University of Lisbon), Professor Tamiru Abiye (University of Witswaterstrand, South Africa), Dr. Jude Cobbing (formerly at the British Geological Survey, a researcher and consultant in South Africa), Dr. Omvir Singh (Department of Geography, Kurukshetra University, India), Dr. L. Suri Naidu, (International Water Management Institute, India), Professor Jovan Romelić (University of Novi Sad, Serbia), Professor Paul L. Younger (University of Glasgow, UK), Dr. Ismail Yusoff (University of Malaya, Kuala Lumpur), Dr. Matthew Currell (RMIT University, Australia), Dr. Evan P. Dresel (Department of Environment and Primary Industries, Australia), Mrs. Bolajoko Solanke (Australia), Joseph Gurrieri (National Groundwater Program Leader, USDA), Mike Wireman (USA, co-chairman of International Association of Hydrogeologists Network on Groundwater and Ecosystems), and S. Keuning (Bioclear, Groningen, The Netherlands).

It is hoped that this volume will be useful for further collaborative groundwater research and water-related environmental issues for improved assessment and management of water resources.

In: Groundwater
Editor: Segun Michael Adelana
ISBN: 978-1-63321-759-1

Chapter 1

INTRODUCTION TO HYDROGEOCHEMISTRY, ENVIRONMENTAL IMPACTS AND MANAGEMENT PRACTICES

S. Michael Adelana
Agriculture Research, Department of Environment and Primary Industries, DEPI Bendigo Centre, Victoria, Australia

ABSTRACT

Contamination of groundwater sources is a global phenomenon and it's becoming evident that unregulated human activity will lead to further deterioration of the resource in many countries of the world. There is therefore an increasing need for research and informed policies to strategically manage groundwater along with other natural resources. This book has been structured into ten chapters, including an introductory background and contextual material for its contributions, with case studies from US, Canada, Australia, Brazil, Serbia, India, Malaysia, Netherlands and South Africa. Contributions in this book range from hydrogeochemical investigations and evaluation of groundwater usability for water supply as well as for agricultural purposes to groundwater quality descriptions and characterisation. Issues on groundwater overdevelopment and the consequent quality degradation; flow and recharge processes with management implications; groundwater dependent ecosystems and ecological implications are well covered in its content. A number of simulation variants from numerical modelling to characterise groundwater conditions and estimate groundwater resources potential in large basins are presented to support technical decisions and policy making. The prospects and problems of groundwater development are addressed in this book and novel bioremediation technology with more than 10 years of successful application in remediating groundwater contaminated with chlorinated aliphatic hydrocarbons (CAHs) concludes the content. This book will be useful to all within the groundwater community, researchers and policy makers alike.

INTRODUCTION

Groundwater is a valuable resource and would remain essential in meeting rural and urban water supply needs. There is a growing recognition about its valuable and conjunctive use with other water resources for human consumption, agriculture and industrial purposes. The main aim of this publication is to provide selected case studies with methodologies and practical applications to general groundwater quality and management issues rather than a review of basic theoretical concepts and principles available in teaching materials or reference texts. Although the volume can also be useful reference material for scientists and water engineers involved in research but its chapters are more country-specific and its methodologies applicable and related to water resources assessment, development and management.

Over the years, groundwater has been a preferred source for piped water supplies in many urban areas (Howard, 2007; Dillon et al., 2007; Barret, 2004). In addition, poverty reduction and economic growth had driven the development of groundwater resources across rural Africa (Adelana and MacDonald, 2008), Latin-America and the Caribbean (FAO, 2003; Bocanegra et al., 2002) and Asia (Shah, 2005). Thus, the demand for water, particularly the increasing usage of groundwater, places a greater risk of overuse and contamination in many parts of the world.

The facts remain that the quality of groundwater may change during the cause of its development or may become contaminated by anthropogenic activities of which the impacts are not readily evident. In Europe, many cities have abandoned their urban boreholes in favour of reservoirs and rural boreholes, and had to tackle the issues of rising water tables (Lerner, 2004) as too little water had been withdrawn from the groundwater resource. Conversely, groundwater overdrafts are reported today in many regions of the world (e.g. Harou and Lund, 2007; Goyal and Narula, in this volume); and where there are evidences of groundwater mining the same consequence can result tomorrow if there are no equitable management strategies.

On the other hand, contamination of groundwater sources is a global phenomenon and it is becoming evident that unregulated human activity will lead to further contamination of the resource. For example, barium is indicated as generally higher in waters sampled during the rainy period in Barreiro area (Brazil), defining a contamination plume of $BaCl_2$ in the zone of the effluents dam due to niobium mining activities. This is accompanied by high phosphate levels from the effluents (Mancini and Bonotto, this volume). There are water-related health problems resulting from high arsenic and fluoride levels in groundwater (Fantong et al., 2010; Howard et al., 2006; BGS, 2001).

In addition to the problems caused by arsenic and fluoride, groundwater salinity (arising from various sources, including lithologies containing evaporite minerals, residual sea water in continental coastal margins and evapotranspiration) will restrict the total usable storage for domestic use and food production, most notably in semi-arid or arid areas (Edmunds, 2012). The vulnerability of Australia's coastal aquifers to seawater intrusion is not only an area of current concern but also an area of increasing future concern (Ivkovic et al., 2012).

Over three decades, nitrate contamination was one of the most widespread groundwater problems worldwide reported in the literature (Dubrovsky et al., 2010; Tredoux and Tama, 2006; Adelana and Olasehinde, 2003; Nolan, 2002; Buzek et al., 1998; Overgaard, 1984; Custudio, 1982; Hill, 1982; Greene, 1978), affecting drinking water supplies or the usability of groundwater for other purposes. Recent trend in nitrate concentrations in UK groundwater are reported as rising at an average 0.3 mg/l/year (Stuart et al., 2007). Examples of such effects and pollution potentials are still been raised today, and are reported in this volume (Adelana et al., this volume). While even large cities and rural settings are affected, nitrate has also been reported as affecting many groundwater-dependent ecosystems in the United States and the rest of the world (Kreamer et al., this volume).

Environmental Impacts and Challenges

Water Resources management would benefit immensely if focused on the current situation of meeting the drinking water supply needs and the constraints that might have obstructed the sector (Water and Sanitation) from achieving the Millennium Development Goals (MDG) for sanitation. The world has met the MDG target of halving the proportion of people without sustainable access to safe drinking water, well in advance of the MDG 2015 deadline, according to a report issued by UNICEF and WHO in March 2012. This implies that between 1990 and 2010, over two billion people gained access to improved drinking water sources, such as piped supplies and protected wells (WHO/UNICEF, 2012).

Based on the results of these studies on achieving the MDGs for water and sanitation, the aspects of water supply and sanitation in many parts of the world are still far from being a reality. At least 11% of the world's population - 783 million people - are still without access to safe drinking water, and billions without sanitation facilities (WHO/UNICEF, 2012). In many parts of Africa and Latin America, only two thirds of the population have access to safe water within the distance specified by the MDG as 1000 meters from their homes. This affects groundwater quality in these areas, and poses enormous challenge to sustainable management. For example, groundwater studies in several countries of the world show that the contamination of aquifers utilised for water supply is mainly due to improper placement of land-based activities, mostly agriculture and waste disposal including excreta (Abiye et al., 2008; Adelana et al., 2008; Nkhuwa, 2006; Usher et al., 2004).

Despite these obvious needs and inherent challenges, however, sufficient attention has not been paid to the systematic gathering of information on groundwater quality, with the result that data and knowledge about it becomes limited and investment/funding is left to emergency response to sudden outbreaks or occurrences. Therefore, managing groundwater resources for water security and sustainability are hampered because effective management and strategic planning are better informed by research studies built on better monitored data.

This volume highlights the issues and complexities of groundwater quality and environmental impacts surrounding the development and management of the resource. It provides a snapshot of groundwater across various countries (e.g. Serbia, India, Malaysia, etc.), with research emphasis in the 21st century. In South Africa and Australia both nitrate contamination and salinity issues have the potential to become large-scale groundwater problems, as such requiring a large-scale source control, which in itself may take several years or decades to manifest or become effective measures. Therefore, potential groundwater reserve may be unusable or out-of-use for longer periods than necessary if there are no periodic/regular quality monitoring.

Providing mechanisms for information and management policies guided by 'resource-directed' knowledge are key roles that would demand the attention of water scientists and the government at local and national levels. International cooperation and joint-efforts may be needed to manage aquifers that cut- across national boundaries, because degradation in certain aquifers could be long-term and in some cases, permanent for nitrate and salinity management strategies.

While natural geochemical processes play an important role in groundwater quality, issues of water quality is also a limiting factor in quantifying usable fresh groundwater storage (Edmunds, 2012). The chapters that follow (in this volume) provide a number of insights, and can be summarised in the following paragraphs:

Chapters range from characterisation of aquifers to issues on groundwater overdevelopment and the consequent quality degradation; flow, recharge, environmental impacts and strategic discussions of the role of groundwater in development and sustainable management. Groundwater from the Werribee Delta aquifer is a viable resource and using isotope geochemistry to identify where potential surface contamination may occur contributed to groundwater mixing, evolution and implications of conjunctive use of surface and ground water resources within the metropolitan city of Melbourne.

Rapid and ultra-large scale development is progressing in the national capital of Kuala Lumpur and the adjoining areas in Selangor State, Malaysia and has somewhat put greater pressure on water demands. To cope with the water deficit in Selangor State, groundwater is considered as the most viable supplementary water source. The results of nine selected scenarios (simulation variants) characterise the groundwater conditions in these areas and potentially supporting technical decisions and policy making in the state.

In several localities, many native species depend on streamflows and wetlands fed by springs and supported by high groundwater tables. The world's multifaceted dependence on groundwater leads to diverse controversies and myths as described in Kreamer et al. (this volume). Cold groundwater in a riverbed in some locations along the Niagara Escarpment, creates suitable conditions for the spawning of brook trout (*Salvelinus fontinalis*). Ecological sampling showed a rich river fauna in this area, mostly associated with pristine water. Spectrocolorimetry of travertine, clays and shale, and electrical impedance of clays were used to gather supporting information.

There are studies that visualize a potential threat to the quantity and quality of groundwater which may cause serious socio-economic consequences in the future. Thus the groundwater resource requires judicious management and regulations to arrest the declining trend of levels (Goyal Naruda, this volume). Dragovic et al. (this volume) reviews the types of groundwater resources in Serbia, their exploitation, the monitoring of their chemical characteristics, environmental impacts and current activities at national level towards sustainable management.

Finally, the solvents tetrachloroethylene (PCE) and trichloroethylene (TCE) have extensively been used in the past by (metal) industries and chemical laundries. Due to improper storage or disposal, significant amounts spilled into the groundwater. PCE, TCE and their subsequent degradation products dichloroethylene (DCE) and vinyl chloride (VC) spread easily in groundwater and are relatively recalcitrant to biodegradation, resulting in large groundwater areas contaminated with chlorinated aliphatic hydrocarbons (CAHs). Due to their toxic and carcinogenic properties, CAHs pose a serious risk to human health. To reduce these risks, and restore the possibilities to use groundwater as a risk-free resource, bioremediation of CAHs using data from over 300 wells in the Netherlands is presented to demonstrate a cost-effective and low environmental impact groundwater remediation application (de Vries and Nipshagen, this volume).

Some Solutions and Prospects

Groundwater problems due to its usage and complexities of human activities (as highlighted above) vary greatly from region to region and approaches to solving them are quite important locally in different geological settings. There is still a great prospect for groundwater in meeting basic human needs and contributing solutions to socio-economic problems in many parts of the world. Strategies for managing groundwater in such environments with huge dependence on groundwater (e.g. India) are being addressed in this book (Goyal and Narula, this volume); and can prevent future consequences if adopted, where there are current evidences of over-abstraction.

Groundwater because of its hidden nature is most difficult to remediate when contaminated; the remediation of hydrocarbons even pose a greater challenge. In evaluating the bioremediation of chloroethylenes, nine TCE-concept sites (de Vries and Nipshagen, this volume) and a database of sampling wells were used to demonstrate that bioremediation is a valuable technique to remediate groundwater contaminated with CAHs within a reasonable timeframe.

The book, therefore, contributes to ongoing management solutions in the literature which include: better public and community control of water utilities; repairing old water systems, reducing agricultural water use through introduction of drip irrigation, increased legislation on polluting the water resource, increasing water conservation measures and focusing additional resources on water quality management. Management measures to ensure continued access to clean, accessible, and abundant water supplies are needed across the regions. Such measures are required to address the growing concerns about groundwater quality and sustainability.

Water resources can only be successfully developed when sound strategies are formulated at national, provincial or local levels (Nonner, 2006). A careful balance between these should lead to the formulation of an optimum water development strategy. In areas with water shortage and water-stressed regions, actions may be taken towards further development of the groundwater system (Adelana et al., this volume). In such instance, a thorough investigation before groundwater development and continuous monitoring of the system's level and quality (chemical behaviour) are indispensable to sustainable management.

The chapters within this volume consistently highlight the need for systematic monitoring data and information on groundwater quality. Groundwater studies in many countries occurred on an ad hoc basis and mostly project-based, where resources have allowed researchers to follow an issue of interest. The limited knowledge or the lack of groundwater quality data on an individual national scale will hamper development and a well-managed use of this resource in many countries. While new solutions are needed to meet future water demands globally, critical data shortfalls exist for sustained groundwater quality monitoring especially in Africa, Asia and Australia. Effective management of the groundwater resource is therefore challenging under this conditions and requires improved monitoring for water security.

REFERENCES

Abiye, T. A. (2008). Urban groundwater and pollution in Addis Ababa, Ethiopia. In: Adelana, S. M.A. and MacDonald, A. M. Applied Groundwater Studies in Africa. Volume 13, CRC Press/Balkema, Leiden.

Adelana, S. M.; Xu, Y.; Vrbka, P. (this volume). Hydrogeochemical characterization of the Cape Flats aquifer (South Africa): Constraints for water supply augmentation. Chapter 3.

Adelana, S. M. et al. (this volume). Flow, recharge and mixing processes in the Werribee Basin (Australia) using natural environmental isotope geochemistry: Implications for water resources management. Chapter 7.

Adelana S. M. A. and Olasehinde P. I. (2003). High nitrate in water supply in Nigeria: Implications for human health. *Water Resour.*, 14(1): 1–11. 22.

Barret, M. H. (2004). Characteristics of urban groundwater. In: Lerner, D. N. (ed.) Urban groundwater pollution. IAH International Contributions to Hydrogeology, 24, Balkema, The Netherlands, 29–51.

Bocanegra, E.; Martinez, D. and Massone, H. (eds). (2002). Groundwater and Human Development, Mar de Plata, Argentina.

BGS and DPHE (2001). Arsenic Contamination of Groundwater in Bangladesh. BGS Technical Report WC/00/19.

Buzek, F., Kadlecov´a, R., and Z´ak, K. (1998).Nitrate pollution of a karstic groundwater system. In: Isotope Techniques in the Study of Environmental Change. IAEA, Vienna, pp. 453–464.

Custodio, E. (1982). Nitrate build-up in Catalonia coastal aquifers. *Memoires, Int. Assoc. Hydrogeol.,* 16(1): 171–181.

de Vries, H. M. T.; Nipshagen, A. A. M. (this volume). Bioremediation of chloroethylenes: An evaluation of groundwater recirculation systems. Chapter 10.

Dillon et al. (2007). Conjunctive use of urban surface and groundwater for improved urban water security. In: Howard, K. (eds). Urban groundwater – Meeting the challenge. Taylor and Francis, London.

Dragović, R. et al. (this volume). Groundwater in Serbia: Resources, Environmental Impacts and Sustainable Management. Chapter 5.

Dubrovsky et al. (2010). The quality of our Nation's waters - Nutrients in the Nation's streams and groundwater, 1992–2004: U.S. Geological Survey Circular 1350.

Edmunds, W. (2012). Limits to the availability of groundwater in Africa. *Environ. Res. Lett.*, 7 021003.

Fantong, W. Y. et al. (2010). Geochemical provenance and spatial distribution of fluoride in groundwater of Mayo Tsanaga River Basin, Far North Region, Cameroon: implications for incidence of fluorosis and optimal consumption dose, J. of Environ. *Geochem and Health*, 32:147–163.

FAO (2003). Unlocking the Water Potential of Agriculture. Food and Agricultural Organisation, Rome.

Fekri, A. Benbouziane, C. Marrakchi and M. Wahbi (2008). Impact of the Casablanca municipal landfill on groundwater resources (Morocco). In: Adelana, S. M. A. and MacDonald, A. M. Applied Groundwater Studies in Africa. Volume 13, CRC Press/Balkema, Leiden.

Goyal S. K. and Narula A. K. (this volume). Overdevelopment of groundwater and effect on its movement and quality in an agriculture dominated area in north-western India. Chapter 4.

Greene, L. A. (1978). Nitrates in water supply abstractions in the Anglian region: Current trends and remedies under investigation. *Water Pollut. Control,* 77(4): 478–491.

Harou, J. J.; Lund, J. R. (2007). Economic and water management effects of a no overdraft policy: California's Tulare basin. The global importance of groundwater in the 21st century: Proceedings of the International Symposium on Groundwater Sustainability, N. Hernandez-Mora, A. de la Hera, and S. Ragone, eds., NGWA.

Hill, A. R. (1982). Nitrate distribution in the ground water of the Alliston region of Ontario, Canada. *Ground Water*, 20(6): 696–702.

Howard, K. (2007). Urban groundwater – Meeting the challenge. Taylor and Francis, London.

Howard, G. et al. (2006). Groundwater and public health (Chapter 1) In: Oliver Schmoll, Guy Howard, John Chilton and Ingrid Chorus (eds.) Protecting groundwater for health: Managing the quality of drinking-water sources, WHO.

Ivkovic et al. (2012). National-scale vulnerability assessment of seawater intrusion: summary report, Waterlines report, National Water Commission, Canberra.

Kreamer, D. K.; Stevens, L. E.; Ledbetter, J. D. (this volume). Groundwater Dependent Ecosystems - Science, Challenges, and Policy Directions. Chapter 9.

Lerner, D. N. (2004). Urban groundwater pollution. IAH International Contributions to Hydrogeology, 24, Balkema, The Netherlands.

Mancini L. H.; Bonotto D. M. (this volume). Hydrogeochemistry in groundwaters of Barriro, Araxa City (Brazil). Chapter 2.

Nkhuwa, D. C. W. (2006). Groundwater quality assessments in the John Laing and Misisi areas of Lusaka. In: Xu, Y. and Usher, B. (eds.) Groundwater pollution in Africa, Taylor and Francis, London, 239-251.

Nolan, B.; Hitt, K.; Ruddy, B. (2002). Probability of Nitrate Contamination of Recently Recharged Groundwaters in the Conterminous United States. USGS - Published Research, Paper 58.

Nonner, J. C. (2006). Introduction to hydrogeology. Revised edition, Taylor and Francis, London. 252p.

Overgaard, K. (1984). Trends in nitrate pollution of groundwater in Denmark. *Nordic Hydrol.*, 15(4–5): 174–184.

Shah, T. (2005). Groundwater and Human Development: Challenges and Opportunities in Livelihoods and Environment. *Water Sci. Technol.,* 51(8):27-37.

Stuart, et al. (2007). Screening for long-term trends in groundwater nitrate monitoring data. *Quarterly Journal of Engineering Geology and Hydrogeology,* 40. 361–376.

Suratman, S. (this volume). Sustainable groundwater resources management for Langat Basin, Selangor (Malaysia) through numerical modelling. Chapter 6.

Swatland, H. J. (this volume). Groundwater streams running over clay on the talus slope of the Niagara Escarpment (Canada): an overview with ecological implications. Chapter 8.

Tredoux, G., Tama, A. S. (2006). Nitrate in groundwater in Southern Africa. In: Xu, Y. and Usher, B. (eds.) Groundwater pollution in Africa, Taylor and Francis, London, 15-36.

Usher, B. H et al. (2004). Identification and prioritisation of groundwater contaminants and sources in South Africa's urban catchments. WRC Report No. 1326/1/04.

WHO/UNICEF (2012). Progress on Drinking Water and Sanitation. WHO/UNICEF Joint Monitoring Programme for Water Supply and Sanitation, Report published March 6, 2012.

In: Groundwater
Editor: Segun Michael Adelana
ISBN: 978-1-63321-759-1

Chapter 2

HYDROGEOCHEMISTRY IN GROUNDWATERS OF BARREIRO, ARAXÁ CITY, MINAS GERAIS STATE, BRAZIL

Luis Henrique Mancini and Daniel Marcos Bonotto*
Departamento de Petrologia e Metalogenia,
Universidade Estadual Paulista (UNESP), Câmpus de Rio Claro,
Rio Claro, São Paulo, Brasil

ABSTRACT

Alkaline rocks in southern Brazil vary considerably in age and composition, and have been assembled into distinct geographic provinces that are particularly important due to the presence of economical deposits like the renowned carbonatite intrusion of Araxá. The Araxá complex has been reported in literature as Barreiro and is situated about 6 km from Araxá city, Minas Gerais State. It consists of a circular intrusion, 4.5 km in diameter, with the central part mainly formed by a carbonatite predominantly beforsitic in composition. Additional lithologies include mica-rich rocks, phoscorites and lamprophyres, where all these rock types affect the chemical composition of groundwaters occurring at the Barreiro area. This chapter reports the results of a hydrochemical survey held at the Barreiro area, where several monitoring wells have been sampled in order to provide information on the groundwater quality. Two fieldwork campaigns for groundwater sampling were realized in this

* Corresponding author: danielbonotto@yahoo.com.br.

study, they comprise the wet and dry seasons. The samples were analyzed *in situ* for physical parameters and stored in 1-L polyethylene bottles for chemical analyses. The water table depth, temperature and conductivity readings were performed *in situ* on using YSI probe, whereas the pH and redox potential (Eh) data were acquired by a Digimed portable meter. The water samples for chemical analysis were transported up to LABIDRO-Isotopes and Hydrochemistry Laboratory, Rio Claro city, São Paulo State, Brazil. They were filtered in 0.45 μm Millipore membrane for retaining the suspended solids before characterizing the major compounds in the liquid phase. The Hach DR 2000 spectrophotometer was utilized for yielding data for silica, tannin/lignin, alkalinity (as bicarbonate), magnesium, calcium, barium, nitrate, sulfate and phosphate. The sodium, chloride and fluoride measurements were done by potentiometry through Orion combined electrodes. The insertion of the pH and Eh data in an Eh-pH diagram indicated that the obtained data are inserted within a transition zone, tending to exhibit a reducing character, where a highly reducing value was found in the sampling held in the dry season. It was found inverse Eh-pH relationships in both fieldwork campaigns. Bicarbonate was the dominant anion in the monitoring points, whereas Na and K were the main cations, with Mg predominating in some cases. The barium content was generally higher in waters sampled during the rainy period. The hydrochemical data were subjected to statistical analysis that indicated that barium is inversely related to sulfate in the rainy period.

Keywords: Hydrochemistry; carbonatite; Barreiro area, Araxá city; niobium deposit; phosphate mining

INTRODUCTION

The subsurface consists of an upper zone with soil water and a lower saturated part with groundwater, where they are separated by the groundwater table (Mook, 2000). The depth at which groundwater is encountered depends on topography, subsurface structure and climate. Total storage of groundwater in a layer is determined by the porosity, but the ease at which groundwater can flow depends on a combination of interconnection and size of the pores (Mook, 2000). Residence times of groundwater range between months for shallow drainage systems to thousands of years for deeper systems, and especially in arid areas with low recharge and large turnover time (Mook, 2000). Exceptional cases of fossil or residual groundwater flow can be found in extensive groundwater basins in desert areas (Mook, 2000).

The rock matrices exhibit variable properties to stored water, where sand and gravel or fractured bedrock often constitute good aquifer materials (Castany, 1967). In several areas, the main aquifers are typically unconsolidated alluvium, composed of horizontal layers of alternating coarse and fine materials deposited by water bodies (for instance, rivers and streams) (Castany, 1967). Fractures, faults, cracks and joints are also important rock properties that allow aquifers to conduct and store water.

Brazil owns the highest availability of renewable hydrological resources in world (8,233 km^3/year). However, Brazil is the 26th country in terms of social water availability (48,314 m^3/inhab./year), as a consequence of the great unequal distribution of the hydrological resources in the country (57% is concentrated in northern and portions of northeastern and central western regions, where the population density is low) (Borghetti et al., 2004). This advanced degree of unequal water distribution has drawn special attention to the groundwater resources due to problems related to the interaction between the society and the environment (CPRM, 2013).

The use of groundwater in Brazil has increased mainly in the last decades for industrial purposes, irrigation and human water supply. A more accentuated increase is expected in the next years, as a consequence of the technological advances of the sector coupled to the known advantageous of the groundwater in relation to superficial hydrological resources, as in principle, it is less subject do pollution.

The development of the chemical industry has caused the release of several constituents to water bodies and many countries established standards focusing the drinking water quality (WHO, 2011). The mining activities also constitute another important source of contaminants to groundwater in Brazil (Bonotto, 2010).

This chapter reports the results of a hydrochemical survey held at the Barreiro area, Araxá city, Minas Gerais State, Brazil, where several monitoring wells were sampled in order to provide information on the water quality and differences on the aquifers studied. It has also been evaluated if the anthropogenic activities involving the mining and production of fertilizers and niobium taking place there are causing modification in the groundwater quality.

General Features of the Study Area

Araxá is a city located in southeastern Brazil at 19°35'33"S and 46°56'26"W that is 375 km from Belo Horizonte (W), 830 km from Rio de Janeiro (NW) and 581 km from São Paulo (NW) (Figure 1). The town was named after the Indian tribe "Araxas" who lived there at the time it was first discovered. In the 18th century, the cattle breeders and farmers arrived, because they were attracted by the quality of the fertile soil, climate and mineral waters. In the 60's and 70's, the mining of niobium and phosphate fertilizer expanded and also the industrial activity.

The Barreiro area is situated about 8 km from Araxá city center, where there is the well-known Dona Beja springs, whose waters are highly radioactive. Dona Beja is a nickname of Anna Jacintha de São José that lived in the early years of the 19th century, who became one of the most notable women in Minas Gerais history and also a leading figure in the upper ranges of local society (Dannemann, 2013). In addition to a hydrothermal complex, Barreiro area is also characterized by the presence of phosphate and niobium mining activities that were developed by Bunge (past Serrana Mining Co.) and CBMM (*Companhia Brasileira de Metalurgia e Mineração*) companies, respectively.

The Alto Paranaíba Province in southern Brazil includes the renowned carbonatite intrusion of Araxá, covering approximately 16 km^2, which is in general related to a NW-trending linear structure bordering the São Francisco cratonic area and that is thought to be in evidence since late Precambrian times (Traversa et al., 2001). The Araxá complex was previously reported in literature as Barreiro that is located in a fractured zone tending to a major NW direction (Figure 1). It consists of a circular intrusion, 4.5 km in diameter, with the central part mainly formed by a carbonatite predominantly beforsitic in composition (Traversa et al., 2001). A complex network of carbonatite as concentric and radial dykes quite variable in dimension and also small veins ranging from few millimeters to several centimeters in thickness are present intruding either alkaline or country rocks. Additional lithologies include mica-rich rocks, phoscorites and lamprophyres (Traversa et al., 2001).

The weathering layer resulting from the alteration of the alkaline-carbonatite rocks has exhibited accentuated thicknesses, which is locally higher than 200 meters. The alteration material is primarily enriched in P_2O_5, Nb_2O_5, TiO_2, BaO and REE_2O_3. The distribution of these oxides in the intrusion area is irregular, reflecting the complex lithology of the alkaline association (Braga and Born, 1988). The weathering processes promoted the pyrochlore enrichment in the central portion of the circular structure that originated one of the largest world niobium deposits.

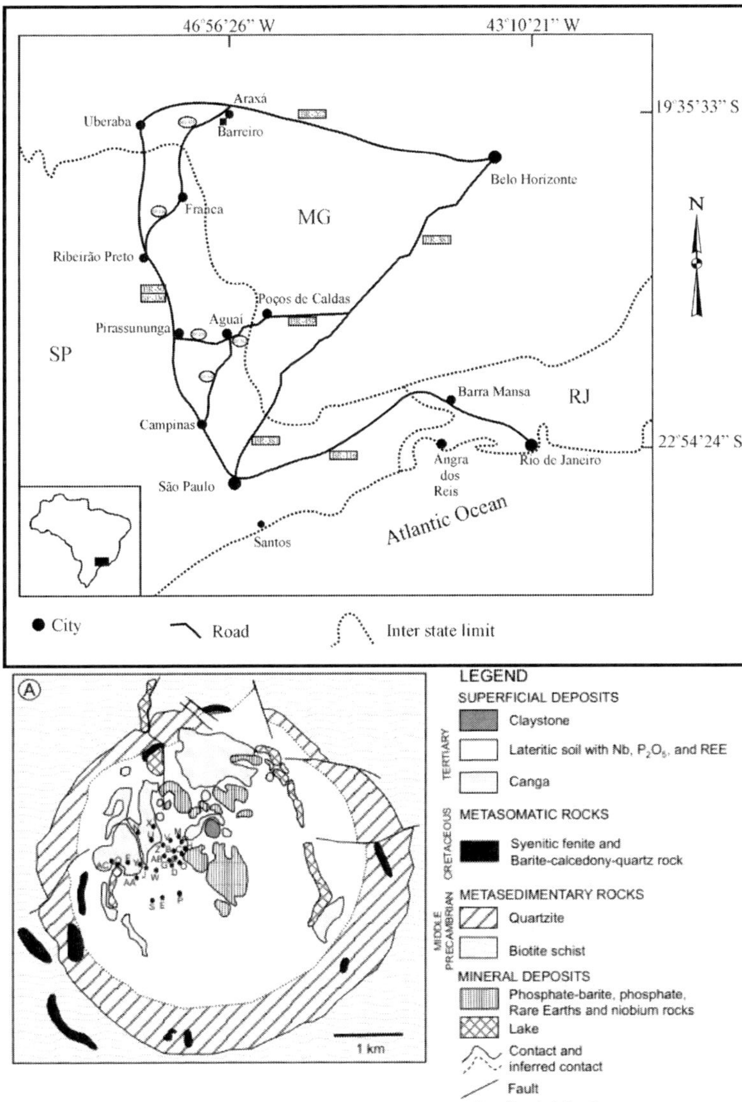

Figure 1. Location of Araxá city and (A) simplified geological map of the Barreiro complex, Minas Gerais State, Brazil, according to DNPM (1987) and Traversa et al. (2001). Samples providing from the boreholes shown in the geological map have been investigated by Traversa et al. (2001).

Pyrochlore is a complex mineral group, whose formula may be expressed by $A_2B_2X_7$. Several changes may occur involving A and B, for instance: A (Na, Ca, Ba, K, U) and B (Nb, Ta, Ti). Barium-pyrochlore is abundant in Araxá, where Ba and Nb dominate the A and B positions, respectively (Braga and Born, 1988). Studies held by IPR (*Instituto de Pesquisas Radioativas*, Belo Horizonte, Minas Gerais State, Brazil) have indicated different pyrochlore generations responsible by the uranium presence (Castro and Souza, 1970). Thus, this radioelement can occur widely disseminated in the mineral matrix as a trace constituent (100 ppm U_3O_8) or may characterize an uraniferous pyrochlore with the U_3O_8 concentration reaching up to 2%.

Groundwater at Barreiro Area

The inauguration of the "Grand Hotel Hydrothermal Complex" in 1944 at the Barreiro area was linked to the existence of two major springs having different hydrochemical composition: Dona Beja and Andrade Júnior.

In fact, Dona Beja spring corresponds to the first mineral exploitation mark at the Barreiro area. These exploited mineral waters are important natural and historic patrimony with significant economic and touristic impacts to Araxá city. The water table interception at the terrain surface due to the topography causes the spring seepage. The water has low salinity but high radioactivity (Lopes, 1956) and the discharge is sustained by the recharging rainwater. Low residence time in the aquifer is expected for these spring waters, where 2-3 km is estimated as the maximum recharge-discharge distance. The aquifer system has been classified as granular, free and semi-confined, mainly in the intrusive body domain (2 km-diameter; variable thickness, reaching up to 200 m at southern portion, close to the niobium mine) (Beato et al., 2000). The aquifer may be locally semi-confined by clayey levels of the weathered layer.

The Andrade Júnior spring is situated in front of the Grand Hotel, possessing a genesis distinct of that of Dona Beja spring. The waters are thermal (33 °C), alkaline (pH=9.6) and sulphurous, exhibiting high salinity (TDS ~ 2 g/L). The water circulation is deep, outcropping in an area where the rock has not suffered significant alteration. According to Beato et al. (2000), the aquifer is fractured, free to semi-confined mainly, occurring in rocks around the carbonatite complex. The recharge is due to direct infiltration of rainwaters in the small areas of outcropping rocks, as well by infiltrating waters in the quartzite residual soil. The groundwater's flow is towards several directions as the ore body is circular and exhibits different altitudes.

Several boreholes were drilled at the Barreiro area during the development of the *Pró-Araxá Project* for investigating the character of the underground flow in the granular aquifer system (FUNTEC, 1984). This aquifer in Barreiro basin is characterized by groundwater flow lines providing upflow from the basin borders and directed downflow to the Waters Park area. The mining activities in the region (phosphate and niobium) modified the groundwater characteristics in the granular aquifer. The niobium mining apparently did not significantly affect the water balance due to its location above the water table. However, its quality changed due to contamination by barium chloride that is a typical residue generated during the Nb separation and extraction. On the other hand, the phosphate mining influenced quantitatively the water balance. Thus, flumes were constructed for regulating it, allowing the groundwater discharge and loss of large water volumes by direct evaporation, which had been previously stored due to the slow movement through the aquifer, underground flow and discharge in downflow areas (FUNTEC, 1984). Table 1 and Figure 2 describe the location of the sampling points for groundwater analysis.

Sampling and Methods

The groundwater sampling points are shown in Figure 2, where most of them are located in the Water Parks and phosphate mining areas. Dona Beja and Andrade Júnior springs were selected for sampling, as well as fourteen boreholes of the *Pró-Araxá Project* and one point in a flume at the phosphate mine area. The borehole samples were collected on using a 1-m long and 50-mm diameter PVC tube containing a ball valve in the lower edge for controlling the water input and storage.

Two fieldwork campaigns for groundwater sampling were realized in this study. The first was performed in the wet period of 16 to 19 February 2000, when boreholes 13sb and 16sb were obstructed and, therefore, were not sampled. The second was realized in the dry season from 12 to 16 September 2000, when boreholes 8sb and 12sb were dry and, therefore, were not sampled. The samples were analyzed *in situ* for physical parameters and stored in 1-L polyethylene bottles for chemical analyses.

Table 1. Location of the groundwater sampling points at Barreiro area, Araxá city, Minas Gerais State, Brazil

Sample code	Latitude - Longitude	Description
1sb	19°38'44"S - 46°57'00"W	Andrade Júnior spring
2sb	19°38'45"S - 46°57'03"W	Monitoring well next Andrade Júnior sp.; Pró-Araxá project code 10/39
3sb	19°39'00"S - 46°56'57"W	Monitoring well along the way to Dona Beja sp.; Pró-Araxá project code 19/42
4sb	19°39'59"S - 46°56'50"W	Monitoring well next Dona Beja sp.; Pró-Araxá project code 19/46A
5sb	19°39'59"S - 46°56'50"W	Monitoring well next Dona Beja sp.; Pró-Araxá project code 19/46B
6sb	19°38'46"S - 46°57'04"W	Dona Beja spring
7sb	19°39'59"S - 46°56'52"W	Monitoring well next Dona Beja sp.; Pró-Araxá project code 19.5/44
8sb	19°39'03"S - 46°56'54"W	Monitoring well along the way to Rádio Hotel ruins; Pró-Araxá project code 21/44
9sb	19°39'36"S - 46°57'00"W	Monitoring well at the area of the Rádio Hotel ruins; Pró-Araxá project code 21.7/40.5
10sb	19°38'04"S - 46°56'45"W	Monitoring well at the area of the Rádio Hotel ruins next to BCM dam; Pró-Araxá project code 21.7/49.6
11sb	19°39'06"S - 46°56'46"W	Monitoring well at the area of the Rádio Hotel ruins; Pró-Araxá project code 23/48
12sb	19°39'09"S - 46°56'53"W	Monitoring well at the area of the Rádio Hotel ruins; Pró-Araxá project code 25/45
13sb	19°39'09"S - 46°56'44"W	Monitoring well at the area of the Rádio Hotel ruins; Pró-Araxá project code 25/50
14sb	19°39'15"S - 46°57'03"W	Monitoring well at the area of the Rádio Hotel ruins; Pró-Araxá project code 28.5/39
15sb	19°39'33"S - 46°57'22"W	Monitoring well at the area of Serrana Mining Co.; Pró-Araxá project code 39.5/28
16sb	19°40'17"S - 46°57'04"W	Monitoring well at the area of Serrana Mining Co.; Pró-Araxá project code 63/39
17sb	19°39'45"S - 46°56'55"W	Groundwater discharging from the phosphate mine of Serrana Mining Co.

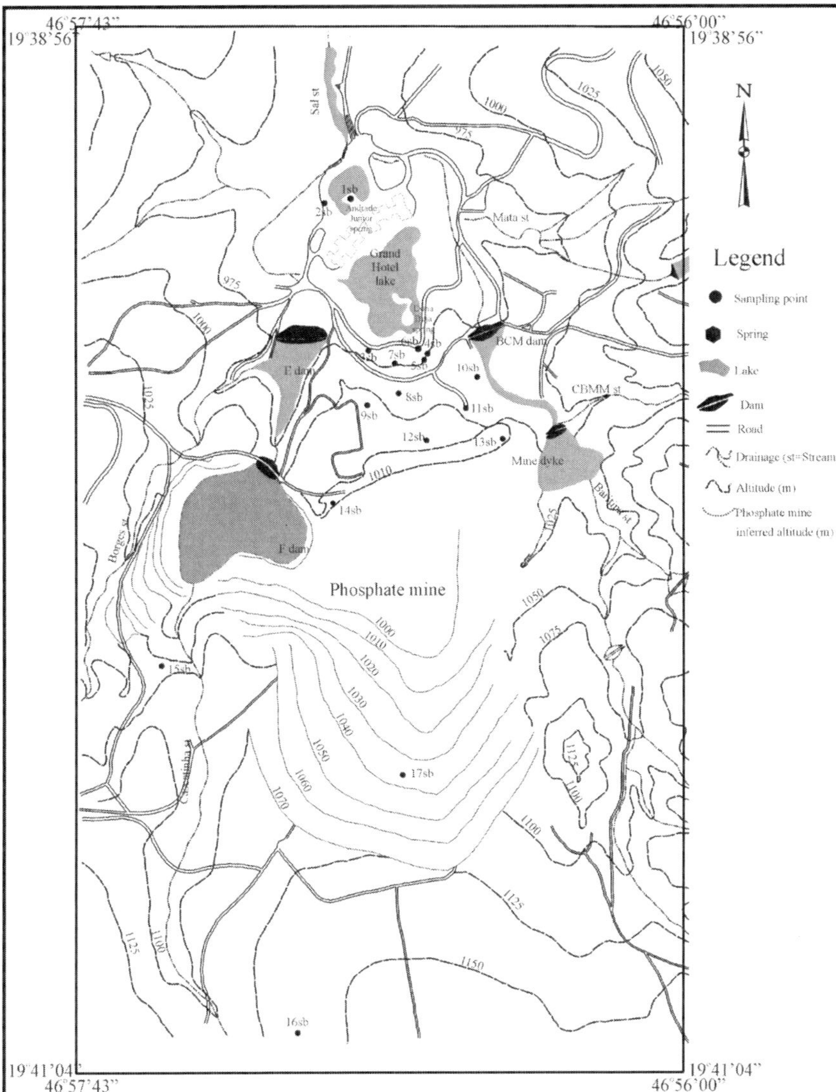

Figure 2. Location of the sampling points for groundwater at Barreiro area, Araxá city, Minas Gerais State, Brazil. Modified from FUNTEC (1984).

The water table depth (NA), temperature and conductivity readings were performed *in situ* using YSI probe model 3000 T-L-C, whereas pH and redox potential (Eh) data were acquired by Digimed portable meter. The water samples for chemical analysis were transported to LABIDRO-Isotopes and

Hydrochemistry Laboratory, Rio Claro city, São Paulo State, Brazil. They were filtered in 0.45 μm Millipore membrane for retaining the suspended solids, whereas the following parameters were characterized in the liquid phase: major cations and anions, silica and tannin/lignin. The Hach DR 2000 spectrophotometer was utilized for yielding data for silica, tannin/lignin, alkalinity (as bicarbonate), magnesium, calcium, barium, nitrate, sulfate and phosphate. The methods adopted are described in HACH (1992). The sodium measurements were done by potentiometry through the Orion Model 9811 combined electrode, coupled to Analion Model IA 601 meter (Mancini, 1997). The anions chloride and fluoride were also evaluated by potentiometry, according to the technique described by (Tonetto, 1996).

RESULTS AND DISCUSSION

Table 2 reports the results of the NA measurements and physical parameters readings, whereas Tables 3 and 4 show the results of the chemical analysis. Figure 3 shows the water table depth (NA) variation according to the wet and dry seasons. The most accentuated differences were verified for sampling points 7sb, 10sb, 14sb and 15sb. However, they were not systematic, as some low values were not found in periods of high rainfall, as expected. It was possible to verify the borehole depth of points, 8sb and 12sb when they were dry (dry season), which corresponded to 20.5 m and 44.2 m depth, respectively.

The NA data in Table 2 allowed estimate of the potentiometric level that was represented in the study area for the two sampling periods (Figures 4-6). Figures 4 and 5 indicate that the flow is concentric towards the Water Parks and also that they do not vary significantly according to the seasonal changes. Figure 6 indicates that the potentiometric level is between 950 and 970 m for most boreholes situated in the Water Parks and at the area of the *Rádio Hotel* ruins. The borehole depth was not available, but it should be between 15 and 50 meters as expected from altitude and NA data. The non-systematic variation of the potentiometric level according to the seasons may be attributed to the water infiltrating from “E”, “F” and “BCM” dams and the phosphate mine dyke.

Table 2. Physical parameters in groundwaters of Barreiro area, Araxá city, Minas Gerais State, Brazil

	Sampling date: 16-19 February 2000					Sampling date: 12-16 September 2000				
Sample Code	NA[1] (m)	Temp. (°C)	Cond. (ms/ cm)	pH	Eh (mV)	NA[1] (m)	Temp. (°C)	Cond. (ms/ cm)	pH	Eh (mV)
1sb	spring	31.2	6.530	9.98	-32	spring	31.4	6.550	10.08	-270
2sb	2.4	24.9	1.220	9.59	45	2.4	23.8	1.560	9.77	-104
3sb	6.7	24.0	0.120	6.83	193	6.7	22.8	0.134	7.00	196
4sb	21.6	22.6	0.170	7.01	186	21.6	22.3	0.178	7.26	162
5sb	18.6	22.5	0.208	7.56	156	18.9	22.3	0.219	7.55	150
6sb	spring	23.4	0.174	7.27	176	spring	22.1	0.195	7.21	165
7sb	6.7	23.4	0.142	6.32	207	5.5	22.5	0.137	6.59	210
8sb	17.7	23.7	0.198	6.54	210	dry	nm	nm	nm	nm
9sb	36.6	22.9	0.049	7.98	124	36.6	22.5	0.051	7.07	101
10sb	4.9	23.3	0.146	6.25	215	6.1	21.4	0.434	6.92	101
11sb	23.2	22.5	0.072	5.69	194	23.5	22.3	0.059	5.88	248
12sb	43.6	22.1	0.178	6.84	140	dry	nm	nm	nm	nm
13sb	obstructed	nm	nm	nm	nm	30.2	22.1	0.080	6.58	69
14sb	28.6	21.9	0.052	5.60	181	26.8	22.5	0.032	5.45	257
15sb	19.8	19.8	0.078	5.51	203	21.3	20.6	0.045	6.09	186
16sb	obstructed	nm	nm	nm	nm	25.3	22.1	0.032	6.23	239
17sb	nm	25.2	0.136	8.05	126	nm	27.8	0.106	7.73	177

[1]NA = water table depth; nm – not measured.

Table 3. Chemical parameters (mg/L) in groundwaters of Barreiro area, Araxá city, Minas Gerais State, Brazil

Sample Code	1sb	2sb	3sb	4sb	5sb	6sb	7sb	8sb	9sb	10sb	11sb	12sb
16-19 February 2000												
Na^+	205.35	6.04	0.22	0.49	0.56	0.45	0.49	0.93	0.81	0.45	1.00	0.75
K^+	1.29	39.20	nm	nm	4.43	4.59	nm	15.00	10.80	4.79	1.40	1.67
Ca^{2+}	0.39	0.40	0.04	0.11	0.05	0.06	<0.02	0.03	<0.02	0.02	<0.02	0.07
Mg^{2+}	0.16	0.62	0.45	0.36	0.35	0.40	0.47	0.37	0.95	0.42	0.78	0.41
Ba^{2+}	1	5	5	5	4	6	4	5	4	6	5	5
HCO_3^-	184	92	78	60	70	64	64	126	38	44	20	58
Cl^-	1.12	20.17	0.25	10.45	13.59	2.62	0.20	0.51	0.25	9.16.	0.98	1.00
F^-	16.77	1.73	0.17	0.30	0.18	0.26	0.26	0.26	0.15	0.50	0.42	0.30
NO_3^-	0.1	0.4	0.3	0.6	0.6	0.5	0.4	0.2	0.3	0.5	1.8	0.7
SO_4^{2-}	200	47	<1	3	2	1	2	<1	2	5	3	3
PO_4^{3-}	1.64	0.58	0.76	0.42	0.35	0.78	0.90	0.41	1.27	1.69	>2.75	0.63
SiO_2	9.9	19.1	16.2	12.1	19.1	20.6	16.2	15.6	7.6	11.1	21.2	28.9
TL^*	<0.05	<0.05	<0.05	<0.05	<0.05	<0.05	<0.05	0.60	<0.05	<0.05	<0.05	<0.05
12-16 September 2000												
Na^+	124.10	45.30	0.30	0.80	0.50	0.30	1.2	nm	0.70	2.10	1.50	nm
K^+	3.06	3.93	1.76	3.13	2.78	4.88	2.90	nm	4.65	4.47	0.58	nm
Ca^{2+}	0.32	0.11	<0.02	<0.02	<0.02	0.02	0.04	nm	<0.02	<0.02	<0.02	nm
Mg^{2+}	0.02	0.14	<0.01	<0.01	<0.01	<0.01	<0.01	nm	<0.01	0.04	<0.01	nm
Ba^{2+}	<1	<1	2	1	2	1	1	nm	<1	3	2	nm
HCO_3^-	804	140	50	64	76	84	68	nm	28	70	10	nm

Sample Code	1sb	2sb	3sb	4sb	5sb	6sb	7sb	8sb	9sb	10sb	11sb	12sb
Cl^-	9.5	20.9	1.5	19.1	21.8	8.8	6.2	nm	3.2	8.0	5.0	nm
F^-	11.2	2.7	0.2	1.3	0.2	0.2	0.4	nm	0.2	0.4	0.3	nm
NO_3^-	<0.1	0.3	0.2	0.4	0.4	0.2	0.2	nm	0.5	1.0	0.9	nm
SO_4^{2-}	220	69	3	4	4	3	8	nm	7	5	2	nm
PO_4^{3-}	1.64	0.73	0.86	0.50	0.45	0.64	1.18	nm	>2.75	>2.75	>2.75	nm
SiO_2	8.1	5.4	9.5	12.8	9.9	20.5	10.0	nm	4.3	24.2	11.2	nm
TL*	<0.05	<0.05	<0.05	<0.05	<0.05	<0.05	<0.05	nm	<0.05	<0.05	<0.05	nm

*TL = tannin and lignin; nm – not measured.

Table 4. Chemical parameters (mg/L) in groundwaters of Barreiro area, Araxá city, Minas Gerais State, Brazil

Sample	16-19 February 2000					12-16 September 2000				
Code	13sb	14sb	15sb	16sb	17sb	13sb	14sb	15sb	16sb	17sb
Na^+	nm	0.49	0.27	nm	0.19	0.20	8.10	0.60	0.60	0.20
K^+	nm	1.12	2.09	nm	3.66	1.12	0.30	0.68	5.56	0.85
Ca^{2+}	nm	<0.02	<0.02	nm	0.03	<0.02	<0.02	<0.02	<0.02	<0.02
Mg^{2+}	nm	0.78	0.33	nm	0.54	0.52	0.92	0.58	<0.01	<0.01
Ba^{2+}	nm	2	5	nm	9	1	1	1	2	<1
HCO_3^-	nm	20	32	nm	60	34	10	16	22	40
Cl^-	nm	0.18	0.08	nm	0.08	0.1	0.2	0.5	0.4	0.3
F^-	nm	0.34	0.11	nm	0.22	0.1	0.2	0.5	0.4	0.3
NO_3^-	nm	0.1	0.3	nm	0.3	0.6	0.1	0.2	1.2	<0.1
SO_4^{2-}	nm	1	1	nm	5	4	1	2	11	3

Table 4. (Continued)

Sample	16-19 February 2000					12-16 September 2000				
Code	13sb	14sb	15sb	16sb	17sb	13sb	14sb	15sb	16sb	17sb
PO_4^{3-}	nm	>2.75	0.03	nm	0.47	0.61	2.74	0.81	0.26	1.57
SiO_2	nm	14.7	14.4	nm	21.3	18.4	11.2	12.3	7.8	6.6
TL*	nm	<0.05	0.70	nm	<0.05	<0.05	<0.05	<0.05	<0.05	<0.05

*TL = tannin and lignin; nm – not measured.

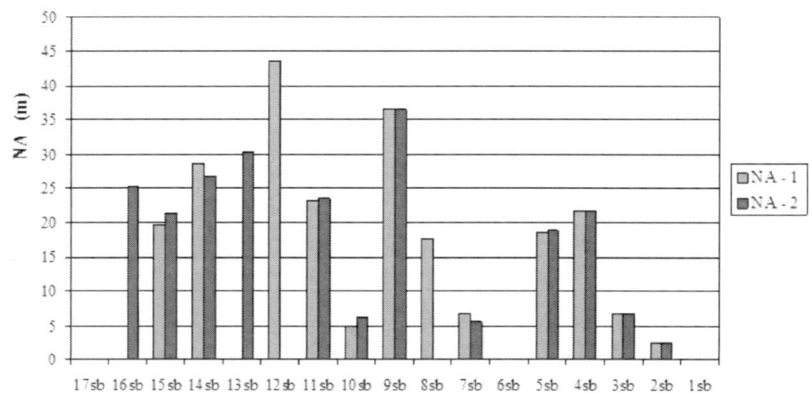

Figure 3. The water table depth (NA) in the boreholes drilled at Barreiro area, Araxá city, Minas Gerais State, Brazil. Sampling date: 16-19 February 2000 (NA-1) and 12-16 September 2000 (NA-2).

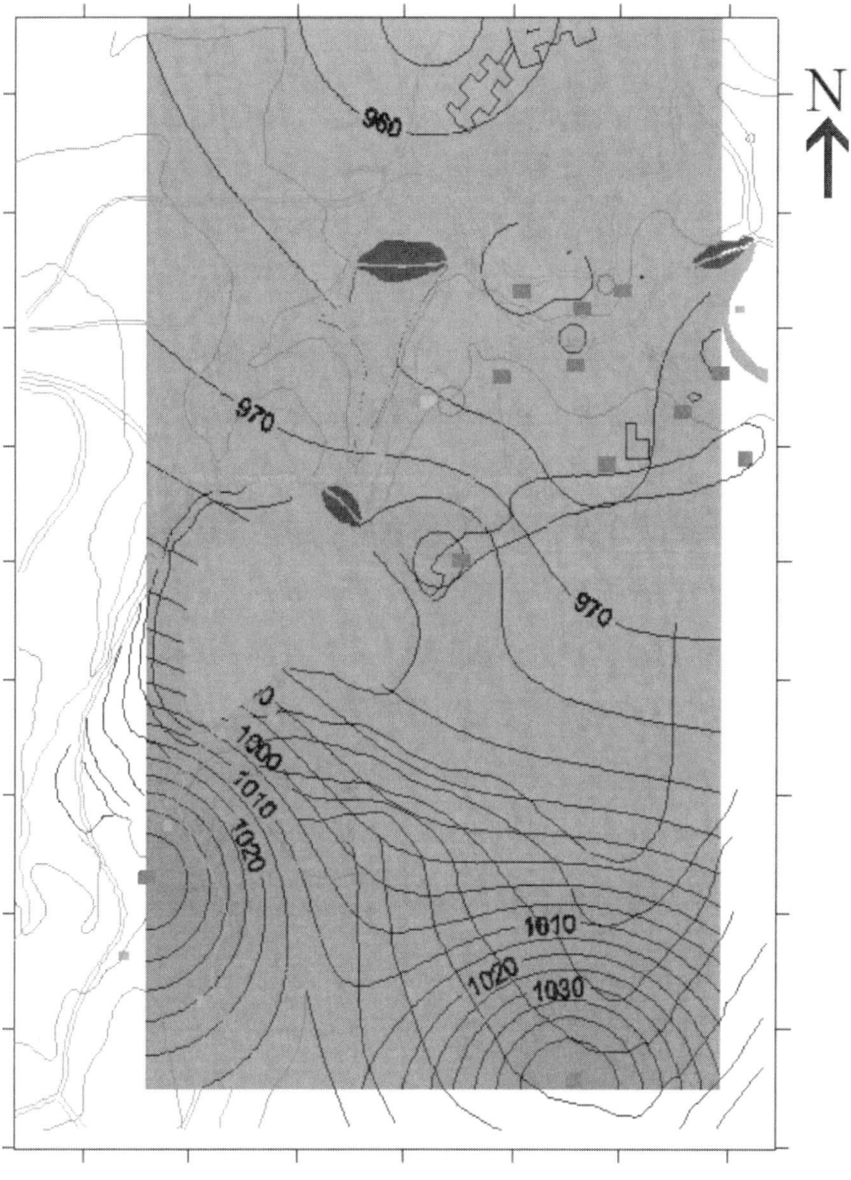

Figure 4. The potentiometric level (m) in the Barreiro area, Araxá city, Minas Gerais State, Brazil. Sampling date: 16-19 February 2000 (PZ-1).

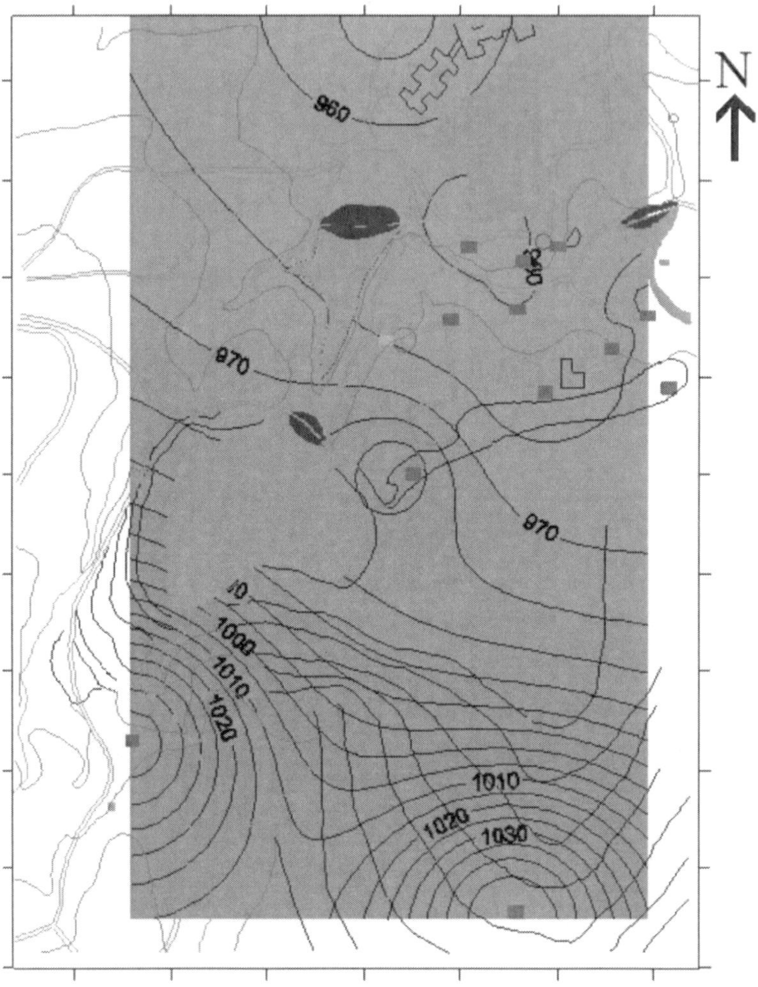

Figure 5. The potentiometric level (m) in the Barreiro area, Araxá city, Minas Gerais State, Brazil. Sampling date: 12-16 September 2000 (PZ-2).

Figures 7 and 8 show the pH and Eh data measured for groundwaters in the two fieldwork campaigns. The values don´t vary significantly for most measurements, suggesting the non-existence of differences among them. The higher amplitude variation was found for Eh data at the close points 1sb and 2sb, possibly due to mixing occurring in the rainy period with more oxidizing waters providing from the shallower groundwater systems.

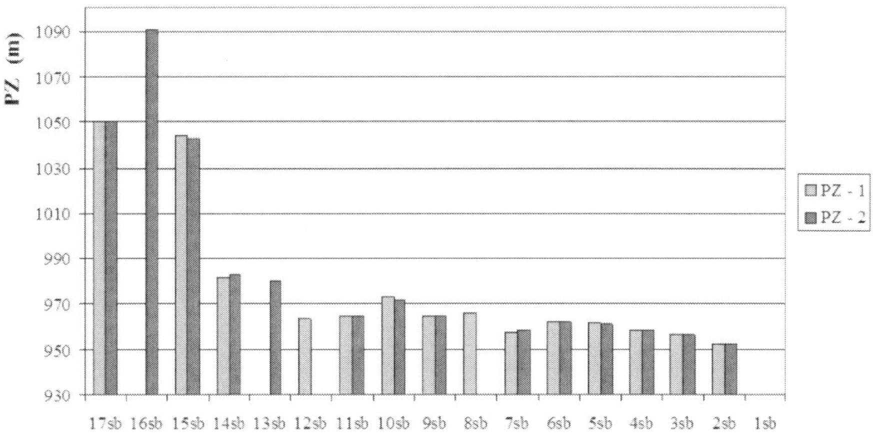

Figure 6. The potentiometric level in the Barreiro area, Araxá city, Minas Gerais State, Brazil. Sampling date: 16-19 February 2000 (PZ-1) and 12-16 September 2000 (PZ-2).

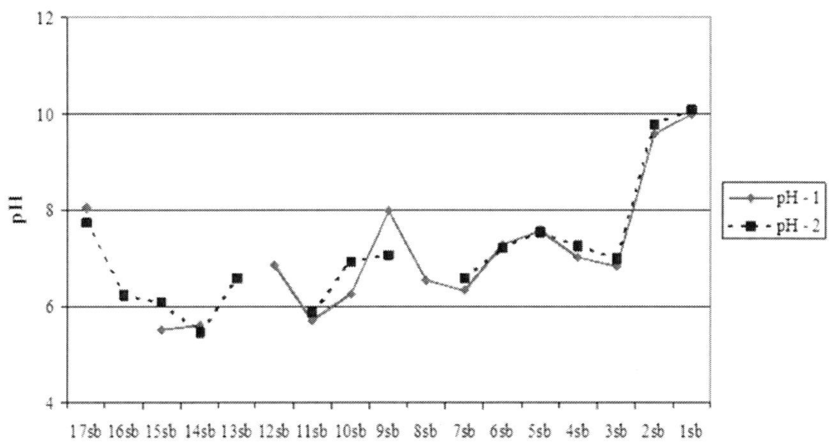

Figure 7. The pH data in groundwaters at Barreiro area, Araxá, Minas Gerais State, Brazil. Sampling date: 16-19 February 2000 (pH-1) and 12-16 September 2000 (pH-2).

The chemical data reported in Tables 3 and 4 are plotted in diagrams shown in Figures 9 and 10. The waters more different are 1sb and 2sb (sulfate-dominated in both campaigns). The point 1sb corresponds to Andrade Júnior spring, whose water circulation is deep in a fractured aquifer occurring around the carbonatite complex. The water in borehole 2sb belongs to the same aquifer system of the Andrade Júnior spring, thus, possessing a similar chemical composition.

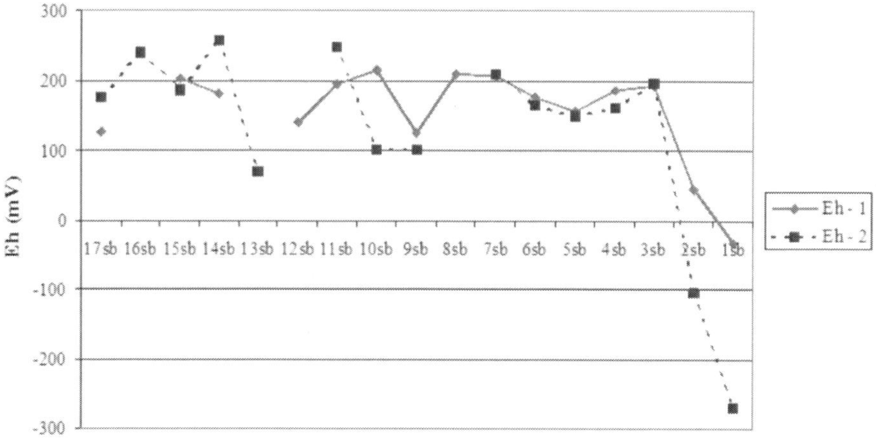

Figure 8. The Eh data (in mV) in groundwaters at Barreiro area, Araxá, Minas Gerais State, Brazil. Sampling date: 16-19 February 2000 (Eh-1) and 12-16 September 2000 (Eh-2).

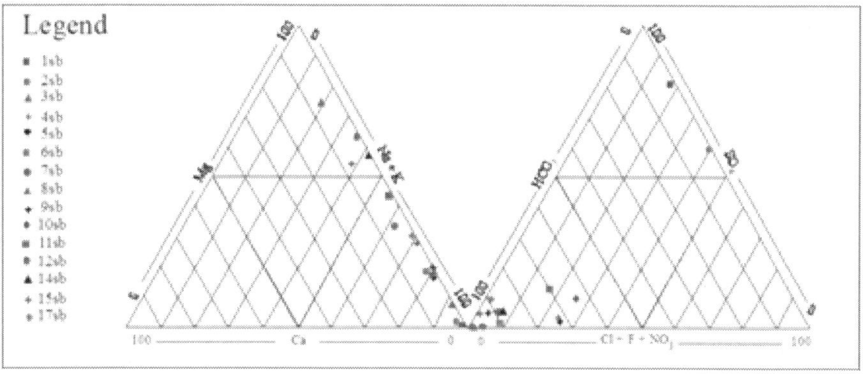

Figure 9. The chemical data in groundwaters at Barreiro area, Araxá, Minas Gerais State, Brazil, plotted in a partial Piper diagram (Piper, 1944). Sampling date: 16-19 February 2000.

Bicarbonate is the dominant anion in the monitoring points, whereas Na and K are the main cations, with Mg predominating in some cases: points 3sb, 4sb, 7sb and 14sb (1st sampling campaign) and points 13sb and 15sb (2nd sampling campaign). Equigranular and fine-grained magnesiocarbonatite is the most abundant carbonatite type occurring in Araxá area, where the carbonate phases consist mainly of dolomite (Traversa et al., 2001). Dissolution processes affecting these rock matrices could justify the Mg and HCO_3^- enhancement in the waters. On the other hand, mining activities of barium-pyrochlore for Nb extraction by CBMM have generated tons of residues directly disposed in the ground, whose leaching can cause the Na and K introduction in groundwater, thus, justifying their dominant presence in several monitoring points.

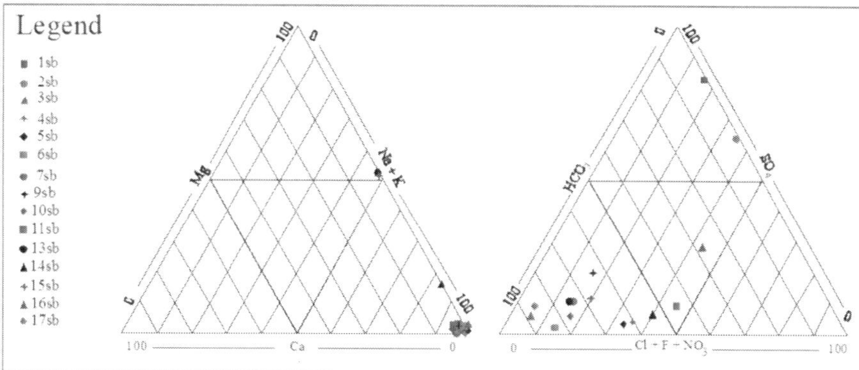

Figure 10. The chemical data in groundwaters at Barreiro area, Araxá, Minas Gerais State, Brazil, plotted in a partial Piper diagram (Piper, 1944). Sampling date: 12-16 September 2000.

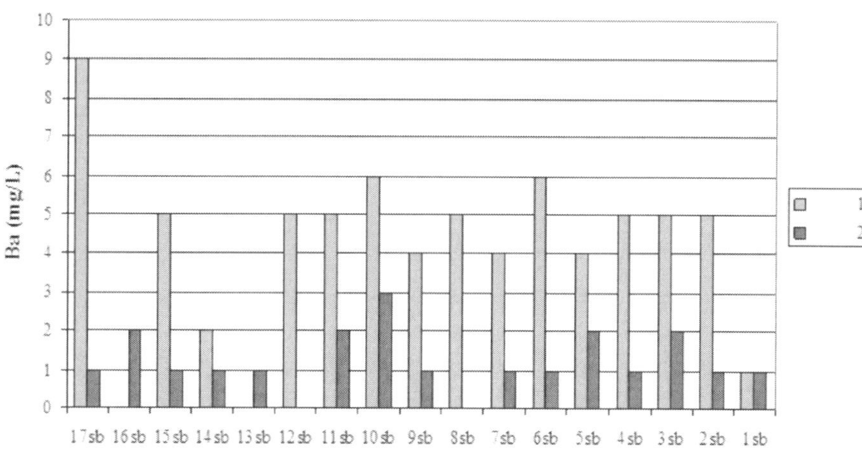

Figure 11. The barium content in groundwaters at Barreiro area, Araxá, Minas Gerais State, Brazil. Sampling date: 16-19 February 2000 (1) and 12-16 September 2000 (2).

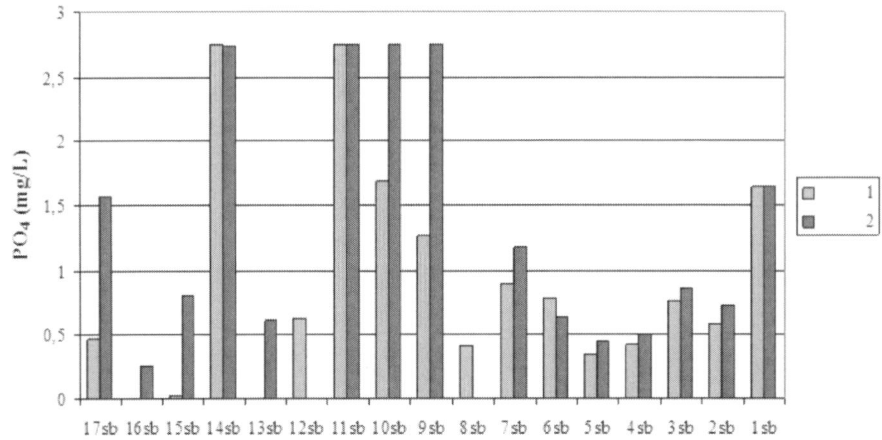

Figure 12. The phosphate content in groundwaters at Barreiro area, Araxá, Minas Gerais State, Brazil. Sampling date: 16-19 February 2000 (1) and 12-16 September 2000 (2).

Figure 11 shows the barium concentration in water samples from the two fieldwork campaigns. The results obtained indicated that Ba is generally higher in waters sampled during the rainy period. This element is widely present in minerals occurring in Barreiro area like barite and pyrochlore, defining a contamination plume of $BaCl_2$ in the zone of the effluents dam due to the niobium mining. Some monitoring points exhibited high phosphate levels (Figure 12) due to the mining of this compound in Barreiro area that takes place at variable depths.

Figures 13 and 14 show the results obtained in the pH and Eh readings during the two fieldwork campaigns. The data are inserted within a transition zone, tending to exhibit a reducing character, where a highly reducing value was found in the second campaign at points, 1sb and 2sb. The parameters pH and Eh reflect, respectively, the protons and electrons activities in the environment, because protons neutralize electrons and vice-versa, the reactions and systems should tend to neutrality under their presence. It has been pointed out by Baas Becking et al. (1960) that four main types of chemical reactions affects the natural systems: a) reactions in which neither protons nor electrons are exchanged, b) reactions involving protons, c) reactions involving electrons and d) reactions in which both protons and electrons are transferred. The last type is more common and, therefore, affects pH and Eh according to the following general trend: lower Eh values tend to occur under higher pH conditions, whereas higher Eh values are contrarily obtained under lower pH conditions (Baas Becking et al., 1960). Thus, the more accentuated characteristics occurring in natural environments would correspond to oxidizing acid and reducing basic systems as often reported in the literature (Baas Becking et al., 1960; Krauskopf and Bird, 1995; Brownlow, 1996). The inverse Eh-pH relationships in Figures 15 and 16 indicate that such trends were in fact verified for the groundwater systems considered in this chapter.

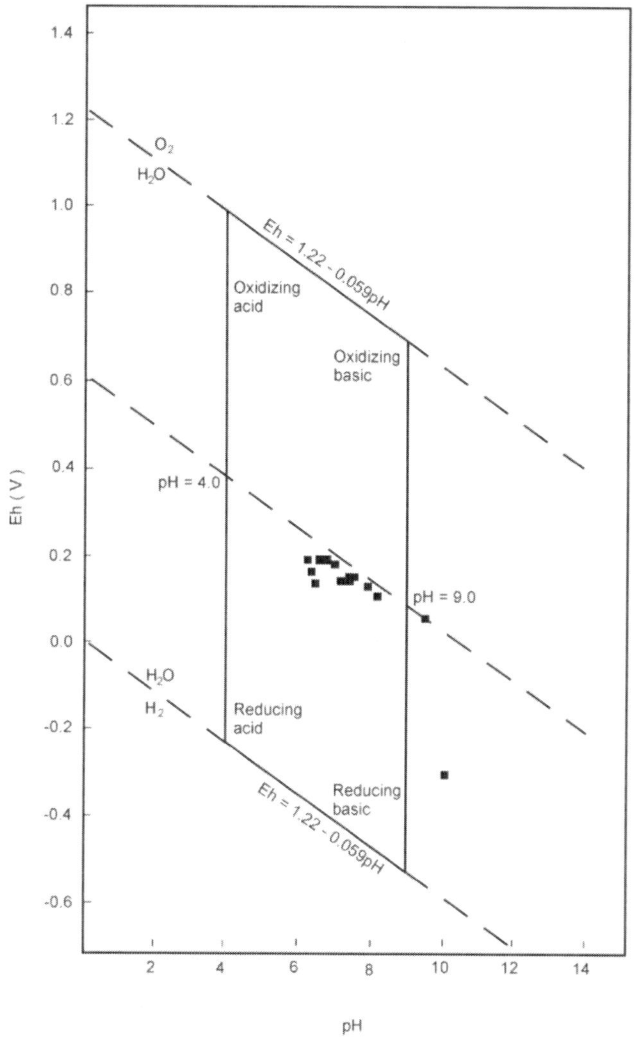

Figure 13. The pH and Eh data in groundwaters at Barreiro area, Araxá, Minas Gerais State, Brazil, plotted in an Eh-pH diagram. Sampling date: 16-19 February 2000.

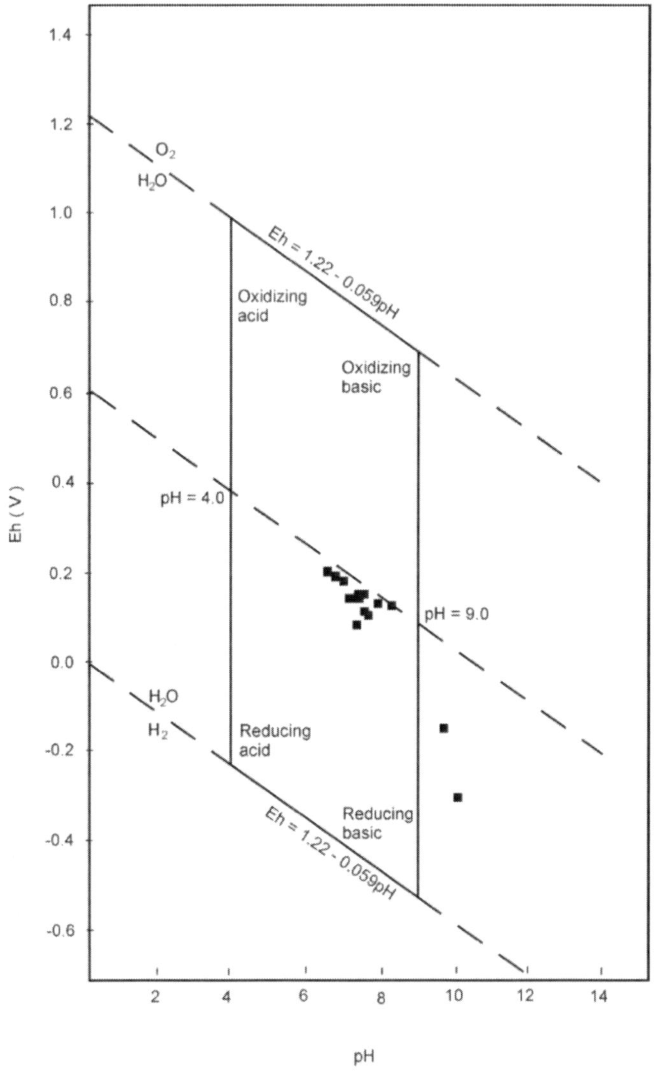

Figure 14. The pH and Eh data in groundwaters at Barreiro area, Araxá, Minas Gerais State, Brazil, plotted in an Eh-pH diagram. Sampling date: 12-16 September 2000.

The hydrochemical data have been subjected to statistical analysis which indicated that barium is inversely related to sulfate in the rainy period (Figure 17). This possibly occurs because contrarily to Ba that is generally higher in the rainy period (Figure 11), dissolved sulfate in groundwater reaches higher values during the dry period. In addition, such trends suggest that the sources of these compounds in groundwater at Barreiro area may be different, perhaps barium-pyrochlore for Ba and pyrite oxidation for sulfate. The pH lowering associated to pyrite oxidation is not clearly verified possibly due to neutralization reactions involving the dissolution of carbonates in the area.

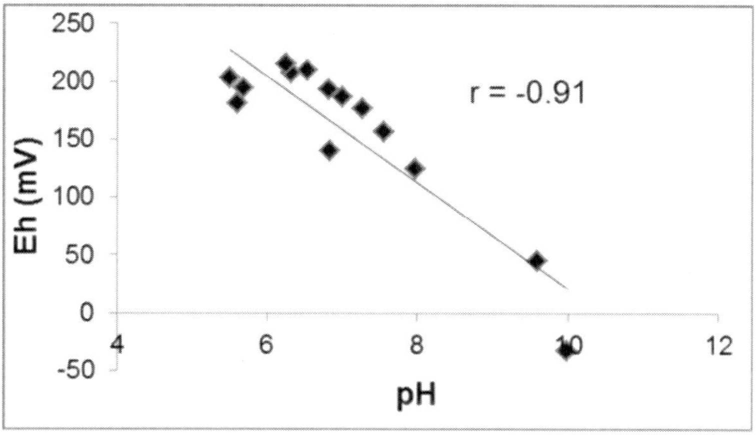

Figure 15. The relationship between pH and Eh in groundwaters at Barreiro area, Araxá, Minas Gerais State, Brazil. Sampling date: 16-19 February 2000.

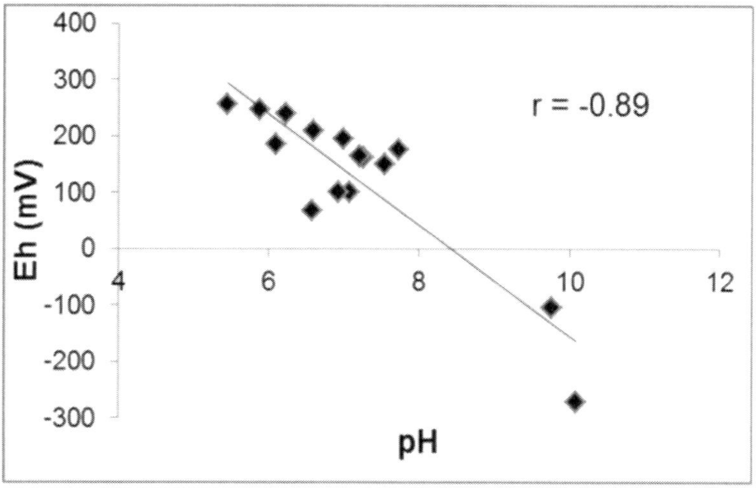

Figure 16. The relationship between pH and Eh in groundwaters at Barreiro area, Araxá, Minas Gerais State, Brazil. Sampling date: 12-16 September 2000.

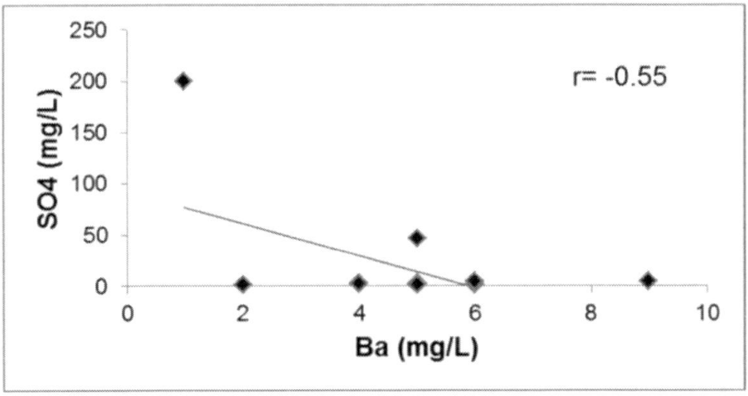

Figure 17. The relationship between barium and sulfate in groundwaters at Barreiro area, Araxá, Minas Gerais State, Brazil. Sampling date: 16-19 February 2000.

Conclusion

The hydrochemical features in groundwater occurring at Barreiro area suffer the influence of mining activities taking place there. They affect the contamination plume eastwards as well the potentiometric level, implying on difficulties for a clear identification of a dominant constituent or its behavior in solution. The Rule No. 2914 of the Health Ministry in Brazil published in 12 December 2011 established maximum permissible concentration in drinking water for several dissolved constituents, where the guideline reference value for barium is 0.7 mg/L. The unique groundwater in the study area sometimes used for drinking purposes is that discharging from Andrade Júnior and Dona Beja springs and the Ba data reported in this chapter indicates that its concentration may exceed the limiting value of 0.7 mg/L. Barite and barium-pyrochlore are abundant minerals in Barreiro area, Araxá city, and the mining activities for niobium extraction generate wastes enriched in barium that may be introduced in groundwater when infiltrating rainwater interacts with them. This could justify the widely spread barium presence in groundwater sampled at the monitoring wells that increases in the rainy season, thus, reinforcing this mechanism as responsible by the contamination plume of $BaCl_2$ in the zone of the effluents dam. Dona Beja and Andrade Júnior springs are an important natural and historic patrimony with significant economic and touristic impacts to Araxá city. The tourism linked to their existence is an activity that has been traditionally conducted along several past decades. The inauguration of the "Grand Hotel Hydrothermal Complex" in 1944 at the Barreiro area was due to the existence of these two major springs. Therefore, its sustainability must be assured by initiatives directed to the groundwater protection. The mining activities in Barreiro area started in times of unawareness about their risks and impacts to the environment and human health. However, such scenario has been gradually modified mainly since "The United Nations Conference on Environment and Development" (UNCED), also known as the Rio Summit, Rio Conference, and Earth Summit (Portuguese: ECO-92) that was a major United Nations conference held in Rio de Janeiro from 3 to 14 June 1992. In UNCED, 172 governments participated (108 sent their heads of state or government), and some 2,400 representatives of non-governmental organizations (NGOs) also participated, with 17,000 people at the parallel NGO "Global Forum", who had Consultative Status. The issues addressed at UNCED included the growing scarcity of water that implies directly on the need to protect the existing hydrological resources, including groundwater. Under this perspective, local and state environmental agencies together with the mining companies at Araxá city have developed efforts in order to mitigate the major hazards caused by the development of their activities. Thus, some initiatives have included the removal of contaminated water and sediments from the small lakes formed by Dona Beja and Andrade Júnior springs, the installation of dams to retain the runoff and groundwater discharge, the construction of drains to change the water flow direction, the re-introduction of vegetal species in order to increase the green area, and the insertion of synthetic geomembrane liners at the bottom of the tailings piles, among other. Despite all, these actions have not solved the pollution problems in Barreiro area yet, on the other hand, they indicate the existence of many people actively engaged in trying to find the best ways to mitigate the environmental damage caused by the uncontrolled mining activities held there in the past. They are positive answers to preserve that natural and historic patrimony for future generations.

ACKNOWLEDGMENTS

FAPESP (Foundation Supporting Research in São Paulo State) and CNPq (National Council for Scientific and Technologic Development) in Brazil are thanked for financial support of this study. Prof. José Manuel Marques is also thanked for constructive comments that helped on the readability of this chapter.

References

Baas Becking, L. G. M., Kaplan, I. R. & Moore, D. (1960). *J Geol*, *68*, 243-284.

Beato, D. A. C., Viana, H. S. & Davis, E. G. (2000). In Proc Joint World Congress on Groundwater; ABAS (Associação Brasileira de Águas Subterrâneas); 1; ABAS: Fortaleza, CE, 1-26.

Bonotto, D. M. (2010). The Poços de Caldas hot spot: a big blast for nuclear energy in Brazil; Nova Publishers: New York, NY.

Borghetti, N. R., Borghetti, J. R. & Rosa Filho, E. F. (2004). Aqüífero Guarani: a verdadeira integração dos paises do Mercosul; Maxigráfica: Curitiba, PR.

Braga, J. R. K. & Born, H. (1988). In Proc Congresso Brasileiro de Geologia; SBG (Sociedade Brasileira de Geologia), *35*, SBG: Belém, PA, Vol. *1*, 219-226.

Brownlow, A. H. (1996). Geochemistry, 2nd ed, Prentice-Hall: Englewood Cliffs, NJ, 580 pp.

Castany, G. (1967). Traité pratique des eaux souterraines; Dunod: Paris.

Castro, L. O. & Souza, J. M. (1970). Estudo de urânio e terras raras associadas ao nióbio de Araxá – MG; IPR (Instituto de Pesquisas Radioativas): Belo Horizonte, MG.

CPRM (Brazilian Geological Survey). (2013). SIAGAS-Information System of Groundwaters. http://siagasweb. cprm.gov.br/layout/pesquisacomplexa.php

Dannemann, F. K. (2013). Dona Beja–Araxá-Minas Gerais. http://www.recantodasletras.com.br/resenhas/ 1677087

DNPM (Departamento Nacional de Produção Mineral). (1987). Principais depósitos minerais do Brasil, DNPM: Brasília, DF.

FUNTEC (Fundação Centro Tecnológico de Minas Gerais). (1984). Relatório diagnóstico do conflito ecológico incluindo obras e medidas recomendadas para atenuação do impacto ecológico da mineração ECOS–Geologia, Consultoria e Serviços Ltd: Araxá, MG.

HACH. (1992). Hach Water Analysis Handbook, Hach Co: Loveland, CO.

Krauskopf, K. B. & Bird, D. K. (1995). Introduction to geochemistry; McGraw-Hill Inc.: New York, NY, 647 pp. Lopes, R. S. (1956). Águas Minerais do Brasil (Composição, valor e indicações terapêuticas); Ministério da Agricultura: Rio de Janeiro, RJ.

Mancini, L. H. (1997). O rádio no ambiente hídrico do Morro do Ferro, Poços de Caldas (MG), M.Sc. Dissertation, IGCE-UNESP: Rio Claro, SP.

Mook, W. G. (2000). Environmental isotopes in the hydrological cycle;UNESCO-IAEA: Paris-Vienna.

Piper, A. M. (1944). *Eos Trans AGU*, *25*, 914-928.

Tonetto, E. M. (1996). O tório em águas subterrâneas de Águas da Prata (SP), M.Sc. Dissertation, IGCE-UNESP: Rio Claro, SP.

Traversa, G., Gomes, C. B., Brotzu, P., Buraglini, N., Morbidelli, L., Principato, M. S., Ronca, S. & Ruberti, E. (2001). *An Acad Bras Ci*, *73*, 71-98.

WHO, World Health Organization, (2011). *Guidelines for drinking water quality*, WHO Press: Geneva.

In: Groundwater
Editor: Segun Michael Adelana ISBN: 978-1-63321-759-1

Chapter 3

HYDROGEOCHEMICAL CHARACTERISATION OF GROUNDWATER IN THE CAPE FLATS AREA, SOUTH AFRICA: CONSTRAINTS FOR WATER SUPPLY AUGMENTATION

Segun Michael Adelana[1,2,•], Yongxin Xu[1] and Petr Vrbka[3]
[1]Earth Sciences Department, University of the Western Cape, Bellville, SA
[2]Agriculture Research, Department of Environment and Primary Industries, DEPI Bendigo Centre, Victoria, Australia
[3]Dieburger Str. 108, D-64846 Groß-Zimmern, Germany

ABSTRACT

A study carried out to evaluate hydrogeochemical characteristics of groundwater in the Cape Flats identified the geochemical processes and their relation to groundwater quality. The distribution of salinity, trends in major ions and ionic ratios imply that groundwater chemistry is largely controlled by a combination of surface and near-surface processes. The groundwater is characterised by Ca–Mg–HCO_3, Ca–SO_4 and Na–Cl. There is little difference in chemistry between the older underlying Malmesbury Formation and the Cape Flats nonetheless in comparison with the Cape Granite, the composition (e.g. total dissolved solids) is ten-

• E-mail: Michael.Adelana@depi.vic.gov.au.

times higher. Sources and mechanisms of salinization were investigated using geochemical techniques. The analysis and interpretation of long-term groundwater monitoring data revealed that salinity and nitrate contamination are potential threats to groundwater usability. High levels of chloride concentration in wells within 8 kilometres of the coast of False Bay is contributing to the deterioration of groundwater quality and could limit the quantity of available sustainable groundwater in this area.

Keywords: Groundwater–Cape Town–Water quality–Hydrogeochemical process–Geochemical evolution – Salinization

INTRODUCTION

The development of groundwater resources in the coastal areas is a sensitive issue that require carefully-planned management programs. Groundwater in sandy aquifers, particularly in the coastal plains, could be highly vulnerable to pollution resulting from anthropogenic activities and seawater encroachments. Therefore, the understanding of processes and chemical reactions related to coastal sedimentary settings as well as seawater intrusion and other pollution pathways become important before the development of coastal groundwater.

Seawater intrusion is a common problem in developed coastal aquifer systems (USGS, 2005; 2006; 2007); the need to plan and manage coastal groundwater reserves has led to several seawater intrusion studies (e.g. Pulido-Bosch et al., 1999; Djabri et al., 2003; Green, 2006) and a wide range of management approaches including the use of hydrogeochemical analyses and numerical modelling (e.g. Sadeg and Karahanoglu, 2008; Gaaloul et al., 2012).

Coastal aquifers have been studied in different countries using hydrogeochemical analyses (Giménez and Morell, 1997; Petalas and Diamantis, 1999; Stuyfzand, 1999; Custodio, 2002, Cabrera and Custodio, 2004, Walraevens and Van Camp, 2005). Natural geochemical processes play an important role in groundwater quality in most of these areas. Water quality is also a limiting factor in quantifying usable fresh groundwater storage (Edmunds, 2012).

The effects of seawater encroachment into fresh water have been investigated using ionic ratios (Petalas and Diamantis, 1999; Sanchez et al., 1999, 2002; Vengosh et al., 2002; El Moujabber et al., 2006; Lee and Song, 2006; Petrides and Cartwright, 2006). However, in the Cape Town area the previous studies on the coastal plain sands have shown limited information on the hydrogeochemical characteristics and evolution of groundwater.

In the past groundwater investigations and extraction mostly takes place from deep fractured aquifers in the Table Mountain Group to avoid pollution by surface activities such as agriculture and indiscriminate waste disposals. This study originated from a long-term monitoring to investigate the status of groundwater in the Cape Flats aquifer and its use in the city's water supply augmentation. This study is aimed at defining hydrogeochemical behaviour of the Cape Flats aquifer in relation to other underlying and neighbouring aquifers and to characterize geochemical processes. The study uses long-term monitoring data to trace the historical and current groundwater quality.

THE STUDY AREA

Location and Description

The Cape Flats is located within the boundaries of the City of Cape Town in the Western Cape Province of South Africa and is positioned in the south-western Cape between 33°30′ and 34°10′ south-latitude and between 18°20′ and 19°00′ east-longitude. Generally, the Cape Flats is taken to be an area bounded by the Cape Town-Muizenberg, Cape Town-Bellville-Kraaifontein, Bellville-Kuilsriver-Strand railway lines and the False Bay coast; with a narrow strip of sand along the western coast, extending northwards from Cape Town and Bellville through Blouberstrand up till Atlantis (Figure 1). Presently, most of the area, underlain by the Cape Flats aquifer, is built-up, with a greater proportion of industrial development from the eastern-side of the City to the northern and southern suburbs.

Topography, Climate and Land-Use

The area around the City of Cape Town has a diverse terrain ranging from low-lying sandy plains to the east and southern end and the Table Mountain to the west and north. The large undulating sandy area connecting the hardrock of the Cape Peninsula with the mainland has been described as the Cape Flats (Schalke, 1973; Theron et al., 1992). This area has an average elevation of 30 m above sea level (a.s.l.), although sand dunes are common on the Cape Flats, with the highest elevation of dunes only 65 m a.s.l. (Theron et al., 1992).

The city of Cape Town has a typical Mediterranean climate but the generally mountainous nature of the area results in contrasting rainfall distribution pattern in the region. Annual precipitation varies between 400 mm and 800 mm, with long-term mean of 600 mm/a. The city of Cape Town receives most of its rain in the winter months (May-August) with a relatively drier period (<20 mm/month) from November to March (Figure 2).

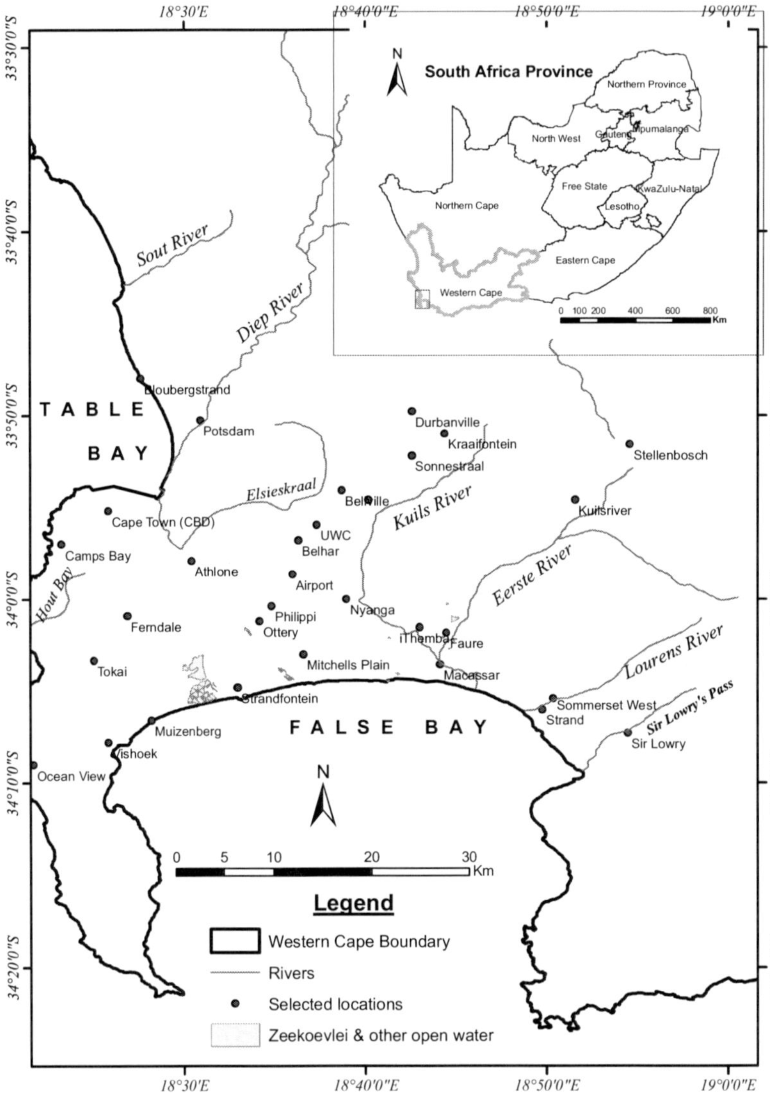

Figure 1. Location of the Cape Flats area in the Western Cape, South Africa.

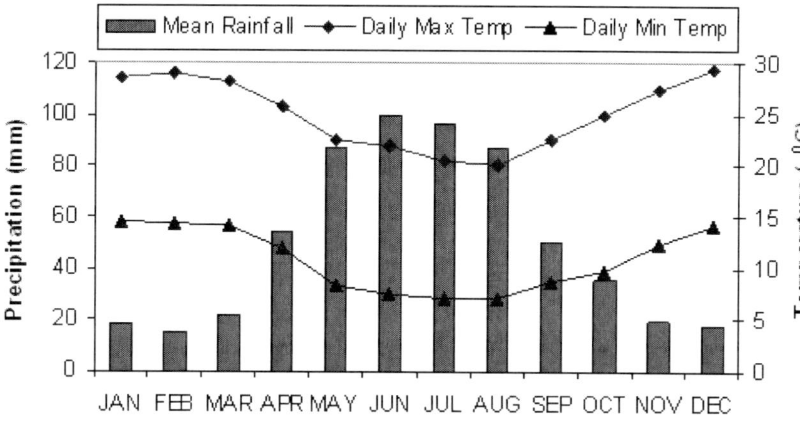

Figure 2. Mean temperature variation with monthly rainfall in Cape Town (1933-2010) Data source: South African Weather Bureau.

The average daily maximum temperature in Cape Town is about 28°C in mid-summer and 17°C in mid-winter. Annual PET, summed from the daily calculations generated by CROPWAT, is higher than AET in order of nearly 2.5 (Adelana, 2011). Land use in the region is dominated by residential/industrial, parks and horticulture (Adelana and Xu, 2008). The Greater Philippi horticultural area, which produces 70-80% of Cape Town's vegetable and fresh produce (such as lettuce, onions, fresh fruit, bananas, potatoes), utilises groundwater from the Cape Flats aquifer for irrigation.

Geology and Hydrogeology of the Cape Flats

The Cenozoic sediments of the Western Cape region are referred to as the *Sandveld Group* (Hendey and Dingle 1983; Malan 1987; Theron et al. 1992). The Cape Flats, which is a component of these Cenozoic sediments, are in places up to 50 m thick (Theron et al. 1992). Figure 3 shows the geological map of the area around the Cape Flats. Geological cross-section A-B-C, shown in Figure 4, illustrates the lithological relationship of the Tertiary sediments in the study area. Part of the area is covered by calcareous sands and limestone deposits while silcrete, marine clays and bottom sediments occur in places (Gerber 1976). The sandy aquifer is underlain by an extensive shaly bedrock aquifer (Malmesbury Shale) (Gerber 1981).

There are reports on the hydrogeology of the area (Henzen, 1973; Gerber, 1976; Vandoolaeghe, 1989).

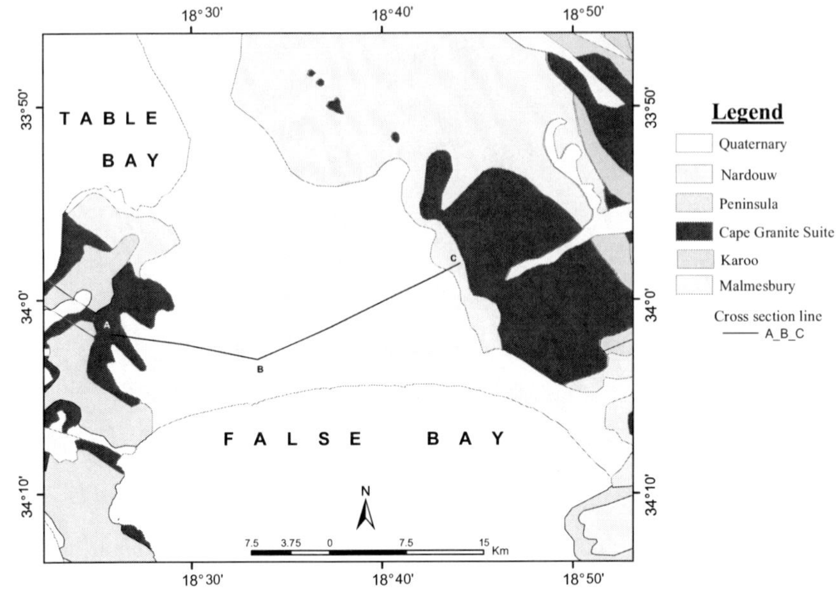

Figure 3. Geological map of the area around the Cape Flats.

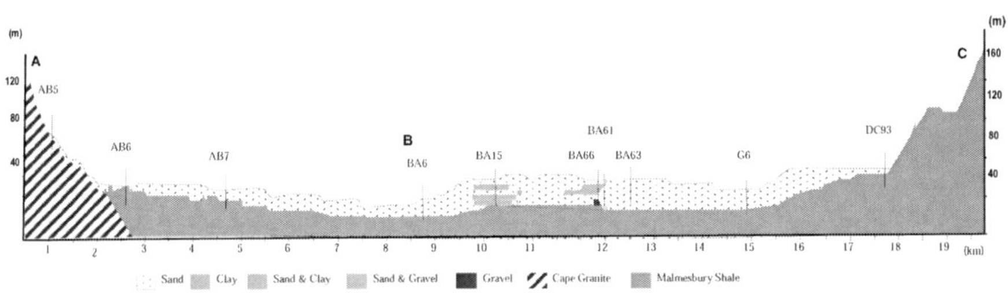

Figure 4. Geological cross-sections A-B-C across the Cape Flats. Locations of some boreholes are shown e.g. BA15.

Updated hydrogeological information on the study area are described in Meyer (2001) and reviewed in Adelana and Xu (2006). Groundwater level range from 0.2 to 20.6 m with a yearly variation between 0.001and 0.2 m (Adelana, 2011).

Yield analysis of about 497 boreholes in the Sandveld Group indicates that 41% of boreholes yield ≤0.5 L/s while 30% yields >2 L/s (Meyer, 2001). Transmissivity values range from 32.5-619 m^2/d (Adelana, 2011), but typical values are between 200 and 350 m^2/d (Gerber, 1981). Recharge rates using Chloride Mass Balance range from 29.0 mm/a (5.5% of rainfall) to 46.6 mm/a

(8.1% of rainfall) for the years 2002-2005; while that from Water Table Fluctuations vary between wet and dry years and between locations, with a range of 17% to 47% of annual precipitation (Adelana et al., 2006).

GROUNDWATER MONITORING AND ANALYTICAL SAMPLING

About 1,928 samples from 170 boreholes, monitored by the Western Cape Department of Water Affairs (DWA), have been used in this study and are available in the National Groundwater Database (NGDB). Most of the boreholes are still active monitoring wells, even though some (which were consistently monitored for groundwater quality in the past) have been abandoned or decommissioned while some are only monitored for groundwater level. A review of the existing groundwater quality monitoring network indicates that the most of sampling programs have been carried out in the active wells, mostly screened in the Cape Flats aquifer (10-30 m below the water table). However, some of the active wells are screened at relatively deeper depths (up to 138 m) in the Malmesbury Shale (weathered shale).

All groundwater quality monitoring data available in the Cape Town area have been compiled (Table 1) and the distribution of wells and sampling points in Figure 5. The sampling intervals differ, some wells are sampled yearly, bi-annually or quarterly, some monthly. This historical data was updated with the (2005-2007) sampling campaigns, which had bi-annual sampling frequency. The dataset is separated into 2 sets with different length of record: a) 1967-2001, and b) 2003-2007.

The pH, electrical conductivity (EC) and total dissolved solids (TDS) were determined at the site using portable field meters and probes during the 2005-2007 sampling campaigns. The carbonate and bicarbonate were determined by acid titration; chloride by $AgNO_3$ titration; sulphate by a titrimetric method using barium perchlorate after passing the samples through cation exchange resin; phosphate by ascorbic acid method using spectrophotometer; calcium and magnesium by EDTA titration; and sodium and potassium by flame photometry in the laboratory of BEMLABS Pty, Somerset West (Cape Town).

The analysed data were processed and interpreted with the programs Hydrogeochemical Analysis Model (Kan et al. 2004) and AquaChem (Waterloo 1999).

Table 1 Summary of the groundwater monitoring network in the Cape Flats area

Monitoring variables	No. of wells (approx.)	Sampling frequency	Data record since (year)
Groundwater level, pH, EC, TDS [a, z]	160	Monthly	1979
Full chemistry (Major, minor ions and trace elements) [b, z]	200	Bi-annual	1967
EC, TDS, Cl, nitrogen (in form of NO_3-N, NH_4-N), phosphate (PO_4), COD/BOD and alkalinity [c]	Several mini-wells and borehole	Bi-annual, differential depth sampling	1986
Routine chemistry [d]	18	4 times per year	1986
Routine chemistry [e]	10	Monthly	1986
Abstraction volume [f]	242	Yearly	1895

[a] Department of Water Affairs (DWA) monthly and bi-annual monitoring
[b] Groundwater chemistry (DWA)
[c] Groundwater monitoring at Coastal Park sanitary landfill
[d] Groundwater chemistry monitored by the Water Division of CSIR
[e] Bellville Solid Waste (record kept by the City of Cape Town)
[f] 140 wells have up-to-date record (WARMS)
[z] Time series for most data not complete; some years are missing, some wells closed

The quality of chemical analysis was checked by ionic balance and the observed charge balance for 1034 groundwater samples is within ±10% (i.e. -7.3% to + 9.3%), which is acceptable for waters of moderate TDS (Eaton et al., 1995). 92% of the total samples fall in the ±5% ionic balance error range while 8% are within the range >5% <10%; and the cation-anion plot gave a confidence level of 98%.

In order to illustrate the ionic composition/classifications, the ions Na^+, K^+, Ca^{++}, Mg^{++}, Cl^-, HCO_3^-, SO_4^{--} and NO_3^- were plotted on a semi-logarithmic Schoeller diagram. This diagram is useful to identify groundwaters with similar ionic concentrations and ratios, comparing them to groundwaters with different ion affinities (Guler et al., 2002). Although the focus of this work is on the Cape Flats aquifer, the ionic composition of the weathered Cape granite and the Malmesbury aquifers, which borders and underlain the Cape Flats in the east and west respectively, were included in the analyses for comparison.

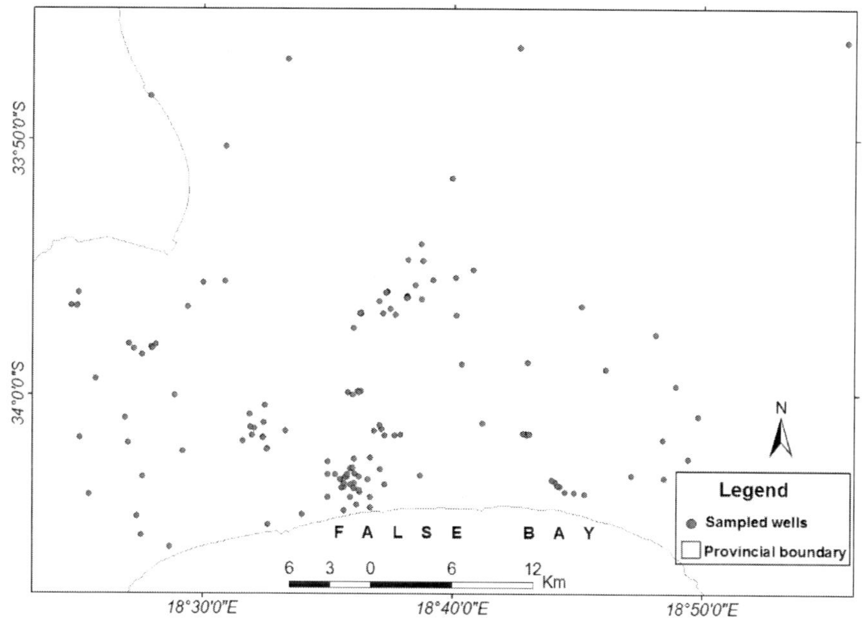

Figure 5. Location of NGDB monitoring wells with chemical data sampled wells, springs and surface water in the current research study.

In summary, the methodology involved analyses and interpretations of several different ionic ratios to: 1) distinguish distributions of chemically similar groundwater and, 2) characterise seawater intrusion and other salinization processes, thus inferring salinity sources. Seawater influence was investigated using ionic ratios. The datasets pertaining to each ionic ratio were partitioned using distinguishing values calculated from inflection points on cumulative frequency curves.

RESULTS AND DISCUSSIONS

Groundwater Hydrogeochemical Composition

The physical and chemical characterisation of the study area is based on the following dataset: 30 samples from wells screened in weathered Malmesbury Shale aquifer, 60 samples from the weathered/fractured Cape Granite and 1,034 samples from the sandy Cape Flats aquifer. The summary statistics of the physical and chemical parameters of groundwater sampled from the Cape Flats aquifer is shown in Table 2. The pH values show a range from 4.8 to 9.1, with a mean of 7.6. The groundwater temperature (from the 2005-2007 sampling) ranges between 17.2°C and 25.4°C.

There is a wide range of values and large standard deviations occur for most of the measured parameters. For example, concentrations of Na^{+}, and Cl^{-} have ranges of 3.3-2,285.8 mg/l and 7.0-5,121.1 mg/l, respectively (Table 2). Ca^{++}, Mg^{++}, and SO_4^{--} also show large variations with ranges of 1.7-366 mg/l, 1.0-321 mg/l and 0.0-846 mg/l, respectively. HCO_3^{-} has a range of 0.1-753 mg/l while NO_3^{-} ranges from 0-248 mg/l. These wide distributions indicate that chemical composition may have been affected by multiple processes. The high concentrations of Na^{+} and Cl^{-} may indicate some level of saline water impact or other salinization process. The concentrations of HCO_3^{-} and Ca^{++} could reflect contribution by water-rock interaction (Park et al., 2005). Despite the wide ranges of values as shown in Table 2, most parameters show log-normal density distributions (Figure 6).

A distorted density distribution can be observed for K^{+} and NO_3^{-}, with concentrations towards low values. Na^{+} and Cl^{-} instead show higher log-values. The distribution helps in explaining the relationships between various measured parameters.

The relationships between various measured parameters are obvious from the correlation coefficients presented in Table 3. Major correlations are:

1) Na^{+}, Cl^{-} and Mg^{++} have a strong correlation (>0.98); and Ca^{++} and SO_4^{--} are correlated with this group (>0.69)
2) Fe^{++}, Na^{+}, Ca^{++}, and Mg^{++} concentrations are correlated (>0.80); Fe^{++}, Mn^{+} are also correlated (0.72) and show good correlation to Na^{+}, Mg^{++}, and Cl^{-} (>0.67)
3) HCO_3^{-} and Ca^{++} have a correlation of 0.67; HCO_3^{-} show weak correlation to other parameters.

The pH and nitrate do not show meaningful correlation with most of the other measured parameters. HCO_3^{-} shows the next lowest correlations with other parameters but correlates to Ca^{++} (0.67). The EC and Cl^{-} were highly positively correlated with most of the other ions (except for K^{+}, HCO_3^{-} and NO_3^{-}), which indicates that these ions are derived from the same source with a limited composition range. These correlations form the basis of discussion for the sources of solutes in groundwater.

Table 2. Statistics of physical and chemical parameters (mg/l) of groundwater in the Cape Flats

Parameter	Minimum	Maximum	Average	Standard Deviation	Number of Samples
pH	4.8	9.1	7.6	0.5	1034
T (°C)	17.2	25.4	19.9	1.5	120
EC (μS/cm)	9.2	4320.0	222.7	408.3	1034
TDS	48.4	10370.0	1181.4	1831.8	1034
Na	3.3	2285.8	231.2	474.7	1034
Ca	1.7	366.0	100.4	75.0	1034
Mg	1	321.3	36.4	72.6	1034
K	0.3	74.9	4.6	7.3	1034
Cl	7.3	5121.1	471.3	1072.3	1034
SO_4	0.3	846.6	70.7	82.5	1034
HCO_3	4	753.0	191.5	102.2	1032
NO_3	0	248.4	2.7	11.0	1034
PO_4	0	11.4	0.1	0.6	1026
NH_4	0	173.0	1.3	9.3	1011
Fe	0	30.3	3.0	6.0	599
Mn	0	2.3	0.1	0.2	591
Ba	0.002	2.0	0.1	0.2	475
Sr	0.001	5.8	1.4	0.8	502
Cu	0	2.7	0.0	0.1	566
Zn	0	1.3	0.0	0.1	540
F	0	3.1	0.2	0.2	1012
Ni	0	0.1	0.0	0.0	510
Rb	0.001	0.0	0.0	0.0	10
B	0.002	3.8	0.1	0.2	556
Al	0.006	1.3	0.8	0.4	476
As	0	1.0	0.1	0.1	493
Be	0.001	0.0	0.00	0.00	377
Cd	0	0.0	0.01	0.00	507
Co	0.001	0.1	0.02	0.01	379
Cr	0	0.1	0.01	0.01	496
Hg	0.002	14.0	0.05	0.77	333

Table 2. (Continued)

Parameter	Minimum	Maximum	Average	Standard Deviation	Number of Samples
Mo	0	0.1	0.01	0.01	453
Pb	0	0.9	0.08	0.10	534
^{2}H (‰)	-22.5	4.0	-13.2	6.2	22
^{18}O (‰)	-4.4	0.1	-2.9	1.0	22

Temperature, T. in C, pH in pH-units, EC in μS cm^{-1},Ca to Al in mg l^{-1}, AS to Zn in μg l^{-1}.

Table 3. Classification according TDS and distribution of groundwater samples in the study area

TDS (mg/l)	Classification	Number and distribution of groundwater samples					
		Cape Flats sand	%	Cape Granite	%	Malmesbury Shale	%
<1,000	Fresh water	758	81.5	60	100	30	50
5,000 - 18,000	Brackish water	100	10.8	-	-	30	50
18,000 -40,000	Saltwater	68	7.3	-	-	-	-
>40,000	Hypersaline	4	0.4	-	-	-	-

Hydrogeochemical Characterisation and Classification

The classification based on TDS (mg/l) and the distribution of the groundwater samples according to the aquifers is shown in Table 3. Seawater influence (18,000-40,000 mg/l TDS) or hypersaline water (TDS >40,000mg/l), was traced in 72 samples from the Cape Flats aquifer only - no such high TDS found in the other two aquifers (Table 3).

The average equivalent concentration of the major ions of groundwaters in the study area is plotted in the Schoeller diagram (Figure 7).

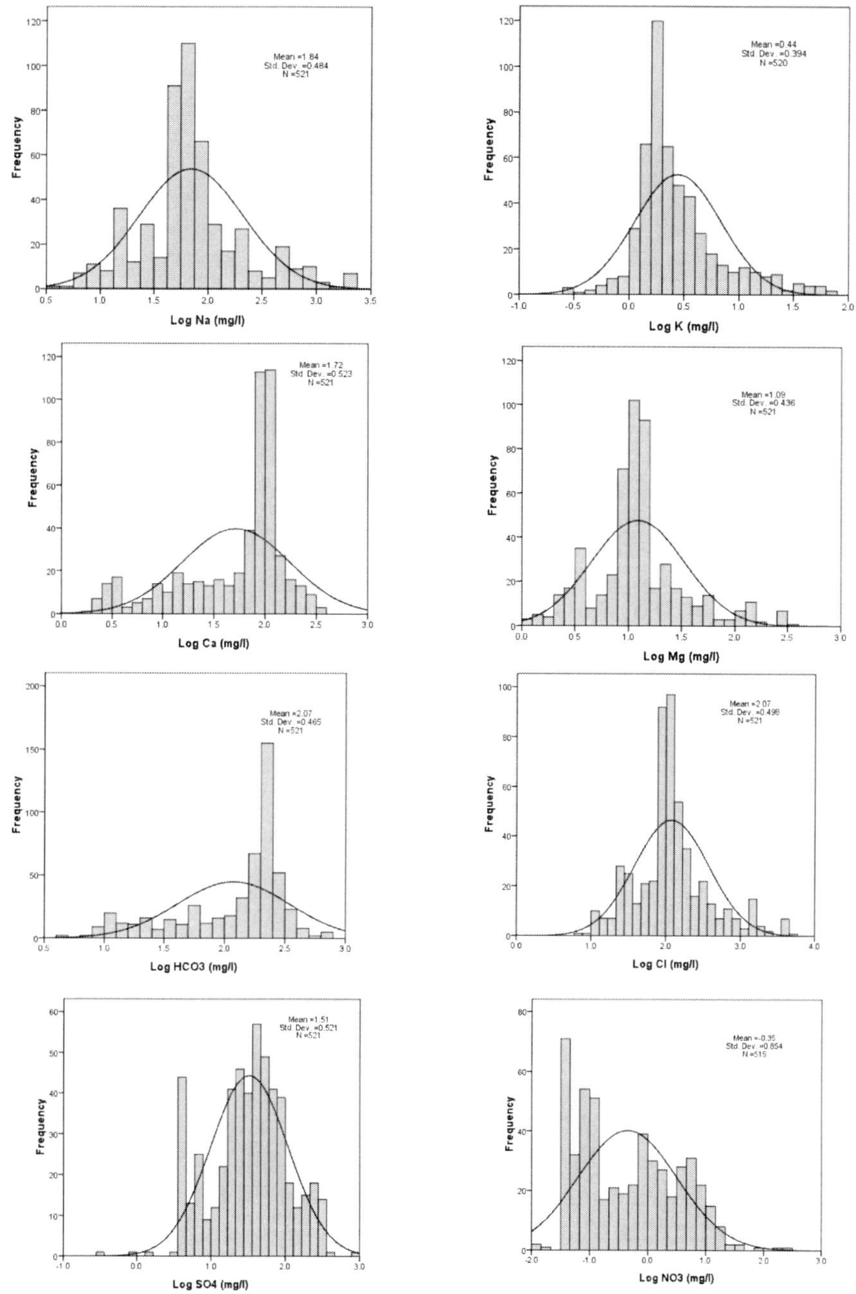

Figure 6. Distribution of major ions in groundwater samples from the Cape Flats aquifer.

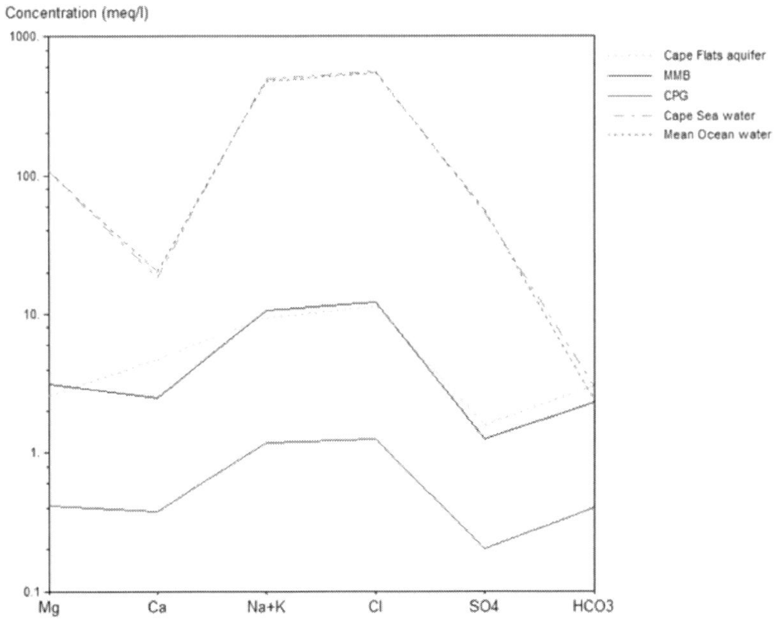

Figure 7. Average equivalent concentrations of major ions in the three aquifers in the area around Cape Town (Symbols: MMB is Malmesbury Shale aquifer; CPG is Cape Granite Aquifer. A plot of the nearby seawater sample (from Saldanha Bay, North-western coastal border of the study area) is used for comparison.

The equivalent concentrations of the elements shown compare well within the three aquifers as indicated by the parallelism of the semi-logarithmic lines. This indicates most of the waters have the similar primary origin (precipitation). The Cape Flats aquifer and Malmesbury Shale are very similar in composition and in concentration as well. Apparently, the Cape Granite shows the lowest ionic concentration whereas the Malmesbury Shale and the Cape Flats have average concentration 10-times higher. The alkaline metals are highest within the Malmesbury Shale, whereas the earth-alkaline metals and nitrate are higher in the Capes Flats groundwater. The variability of major ions could be related mainly to: a) lithological variations promoting ion exchange, b) groundwater mixing, c) impact of farming activities (mainly in the form of fertiliser applications), and d) influence of seawater intrusion (from the coast of False Bay).

Five groups identified from the Cape Flats aquifer illustrate the compositional distinction from the sites of sampling (Figure 8, Table 3). Average concentrations in the groups occur more or less in the same ratio and display similar pattern.

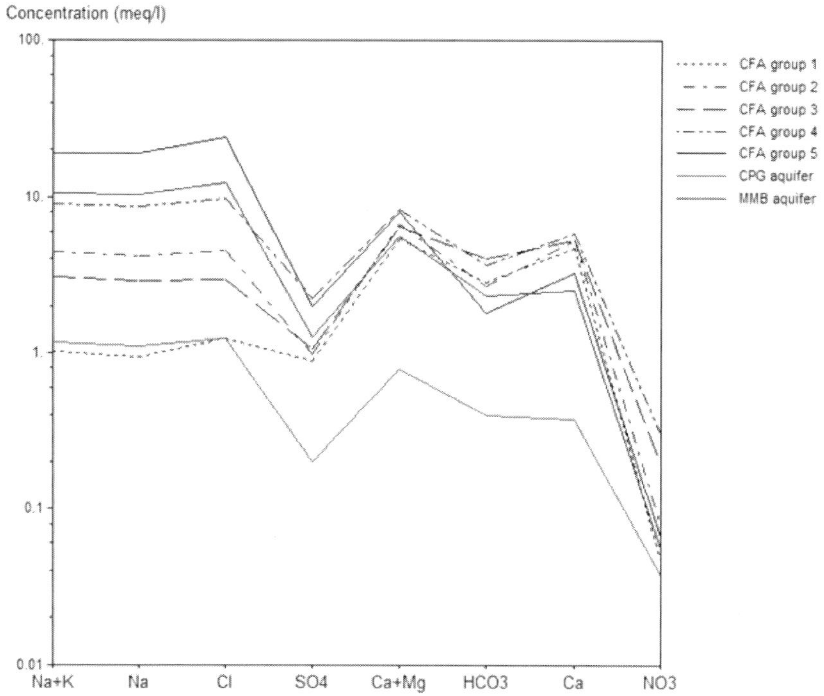

Figure 8. Average equivalent concentrations of major ions and ionic combinations based on the groups 1-5 within Cape Flats Aquifer; compared to Cape Granite and Malmesbury Shale Aquifers.

However, the absolute ionic concentration differs between the groups. Group 1 shows the lowest ionic concentration. Group 5 show the highest values in all ionic concentrations except calcium and bicarbonate. This group is shown to be predominantly Na-Cl type (Figure 8). The transition between these two water types is smooth and continuous suggesting a rather wide transition zone (Figure 9) than a sharp interface between fresh and saltwater.

Based on the cation distribution only (Figure 10), a triangle plot of sodium, calcium and magnesium illustrates the ratio between alkali and alkaline earth elements and also allow the distinction of the groups. Although various hydrogeochemical types were identified, Na-Cl and Ca-HCO_3 appear to be dominant. It is possible that as groundwater travels away from unconfined recharge areas in the Cape Flats towards semi-confined coastal discharge zones it may have evolved through Ca-HCO_3-Cl type via Ca-Na-Cl-HCO_3 type to Na-Cl type; or from Ca-Na-HCO_3-Cl type directly to Na-Cl type or through Na-Ca-Cl-HCO_3 to Na-Cl.

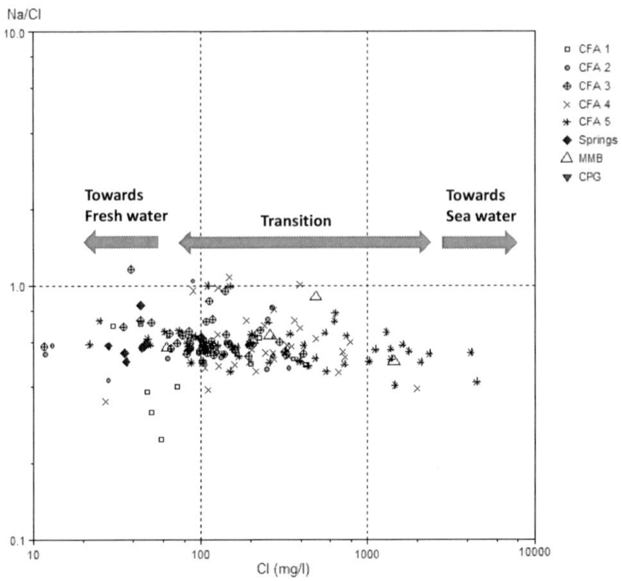

Figure 9. Scatter plot of Na/Cl versus Cl in groundwaters of the study area (symbols as explained in figures 8 and 9).

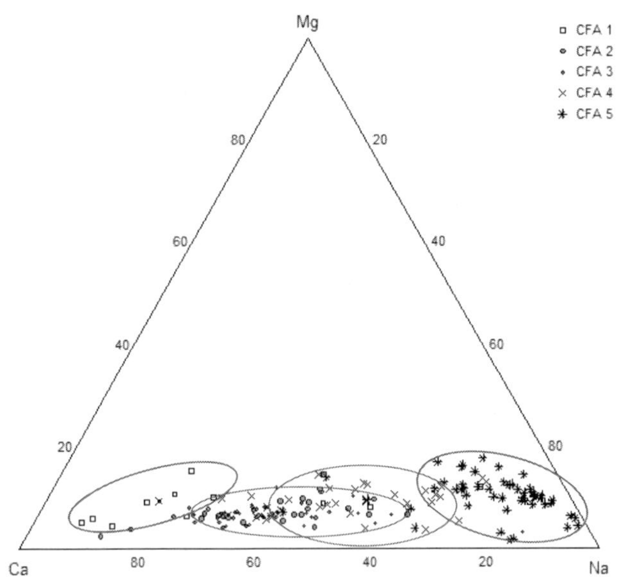

Figure 10. Cation triangle plot of groundwater samples from the Cape Flats aquifer (group 1-5) axis in percentage.

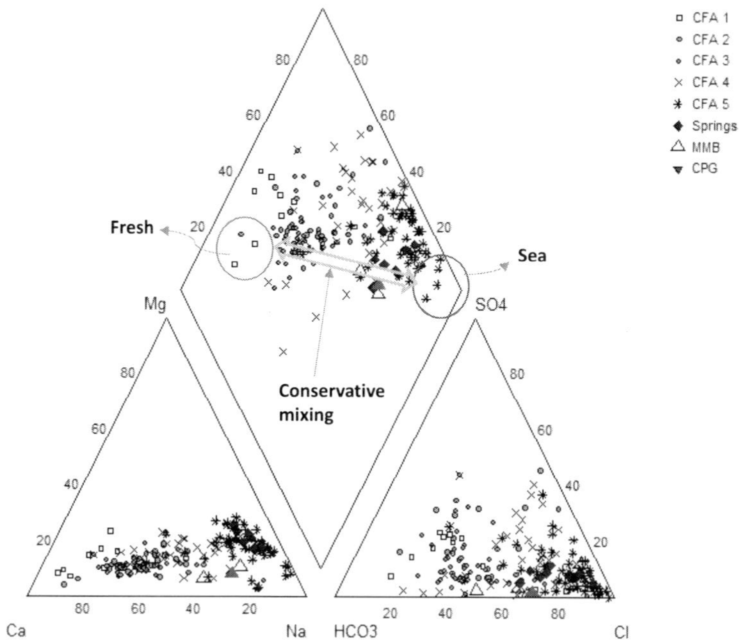

Figure 11. Piper diagram of the Cape Flats Aquifer (CFA) groups 1-5, showing composition of over 900 groundwater samples from different locations and depths in the CFA and representing the transition from fresh- to saltwater, compared to Cape Granite (CPG) and Malmesbury Shale (MMB) aquifers.

These are indications of cation exchange reactions and water mixing, which are prominent processes in groundwater reported elsewhere (Richter et al. 1993; Jeen et al. 2001).

The distinction of the groups is further illustrated in a Piper diagram (Figure 11). In the left cations triangle the Cape Flats groups 1-5 spread across from group 1 with Ca being dominant towards the right side where in group 5 the alkaline metals Na and K dominate the composition. The Cape Granite as well as the Malmesbury Shale aquifer plots here in the right lower corner, indicating Mg being in lower proportions (below 20 meq/l%). In the right-hand anionic triangle the situation is very similar, again showing a development from the Cape Flats group 1 with a high proportion of bicarbonate towards the right and a very high proportion of chloride, >80 meq/l%, in group 5. Generally, seawater is characterised by a high content of chloride. Since the chloride ion does not interact with other ions or the solid phase it can be used as a proxy for seawater. Accordingly, the anion triangle of the Piper diagram gives a good reflection of the seawater intrusion in the Cape

Flats. The water of the Malmesbury Shale aquifer plots close to group 5 while Cape Granite water touches group 4 in the Na+K/Na-Cl end of the plot. Again, the sulphate content is in general low, as indicated by value mainly <20 meq/l%. While all samples are characterised by very low SO_4^{-2} concentrations, HCO_3^- is the dominating anion at lower depths and away from the coast, areas which are flushed by freshwater.

Salinization

The salinity-state of the Cape Flats aquifer is monitored through an observation bore network where electrical conductivity (EC) is the predominant indicator of groundwater quality across the network of bores (Table 1). The EC (μS/cm) is plotted as function of the location distance (km) to the sea (Figure 12). About 16% of the samples from the Cape Flats Aquifer exceed the threshold value (190μS/cm), which differentiates geogenic background from anomalous values (Park et al., 2005; Lee and Song, 2006). Values above the threshold may indicate saline water encroachment, when close to the coast, and/or anthropogenic contamination, least especially for locations farther inland. The highest EC levels, about 10,000μS/cm, are reached within a distance of 8km from the sea (Figure 12). Further inland the values decline and are mainly below the threshold value.

Following the methods of Park et al. (2002) and Lee and Song (2006), the cumulative frequency curves for two parameters (Cl and HCO_3) were used to differentiate 'anomalous' values from 'background' values. Chloride may represent effects of salinisation (or seawater mixing) while bicarbonate is indicative of water-rock interaction. The threshold values were calculated as 327mg/l for Cl and 133.8mg/l for HCO_3 (Figure 13a, b).

Based on the thresholds values, the groundwater in the study area can be divided into four classes (Figure 13c).

(1) 2.4% of the groundwater samples are dominantly affected by the salinisation process;
(2) 59.7% are dominantly influenced by the water-rock interaction;
(3) 7.3% were affected by both water-rock interaction and salinisation;

And

30.6% of the groundwater samples show negligible or low effects by salinisation or water-rock interaction processes.

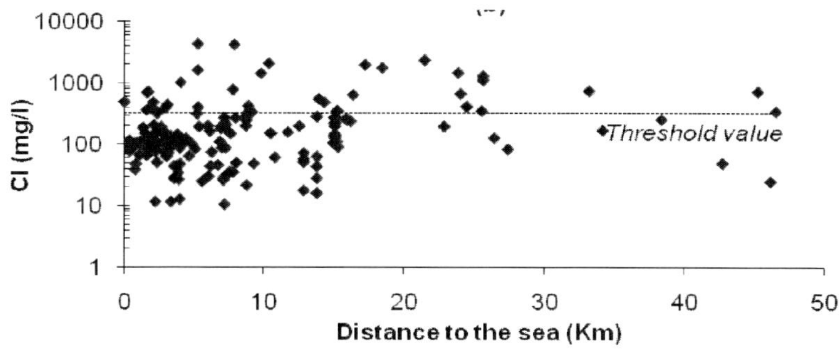

Figure 12. Chloride concentration (mg/l) as function of the location distance (km) to the sea.

Figure 14 shows molar ratios of Na/Cl and SO_4/Cl versus Cl concentrations in groundwater of the study area. Generally, conservative seawater-fresh water mixing is expected to show a linear increase in Na and Cl (Lee and Song, 2006, Sanchez et al., 1999) as well as Na/Cl values. Ratio of values of groundwater less than the seawater ratio 0.86 indicate that fresh groundwaters have been influenced by the saline water. The range of Na/Cl ratios (0.1 to 77) and about 5% were below the seawater ratio and exceeded the threshold value for chloride (Figure 14a). Values close to or clustering around the seawater (ratio) line may indicate recent simple mixing of groundwater with seawater (Mercado, 1985; Lee and Song, 2006).

Similarly, figure 14(b) shows the variations in the molar ratios of SO_4/Cl against Cl concentrations. Molar ratios range between 0.003 and 22.5 while 6.9% of the groundwater samples are less than seawater value of 0.1, indicating influence of seawater. Cl/(HCO_3+CO_3) also indicate seawater influence (Figure 14c). Cl/(HCO_3+CO_3) ranges between 0.16 and 53 and shows a positive linear relation with Cl concentrations, indicating mixing of fresh water with saline water. The classification based on the methods of Lee and Song (2006) and Petrides and Cartwright (2006) shows that Cl/(HCO_3+CO_3 ratios <0.5 are "unaffected" by saline water, while 0.5-6.6 show "slightly/moderately affected", and >6.6 "strongly affected". The ratio further shows that 4.8% of the samples are below the seawater ratio in figure 14c; these are those classified as strongly influenced while another 3.8% are slightly/moderately affected by seawater. In essence, <10% of all the samples analysed have been subjected to seawater influence.

Evapotranspiration (ET) of rainwater in the unsaturated zone prior to recharge (or during recharge) and from shallow water tables (after recharge) is another possibility for higher salinity in the shallow groundwater.

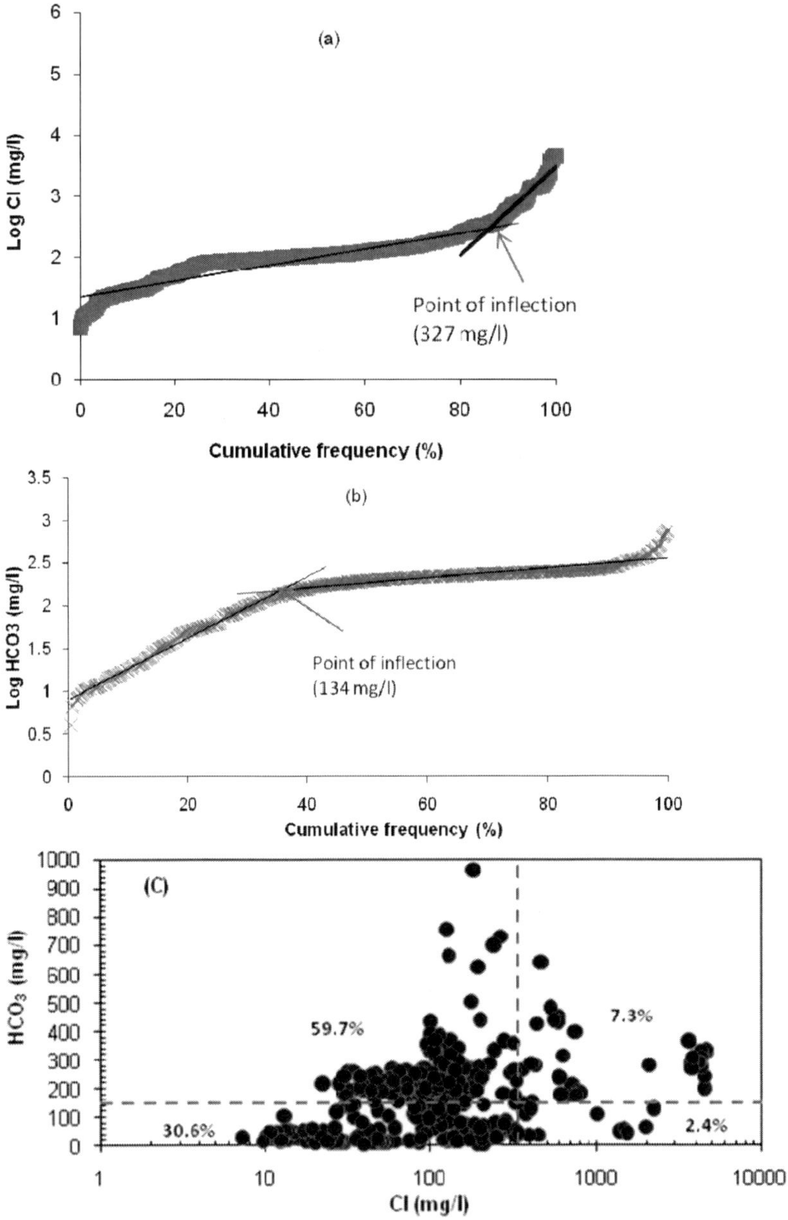

Figure 13. Cumulative frequency curves for: (a) Cl and (b) HCO_3. The inflection points are calculated based on Park et al (2002); Lee & Song (2006). Estimated threshold values have been used to differentiate background from anomalous values. Based on the threshold values, classification of four groundwater types is also shown in (c).

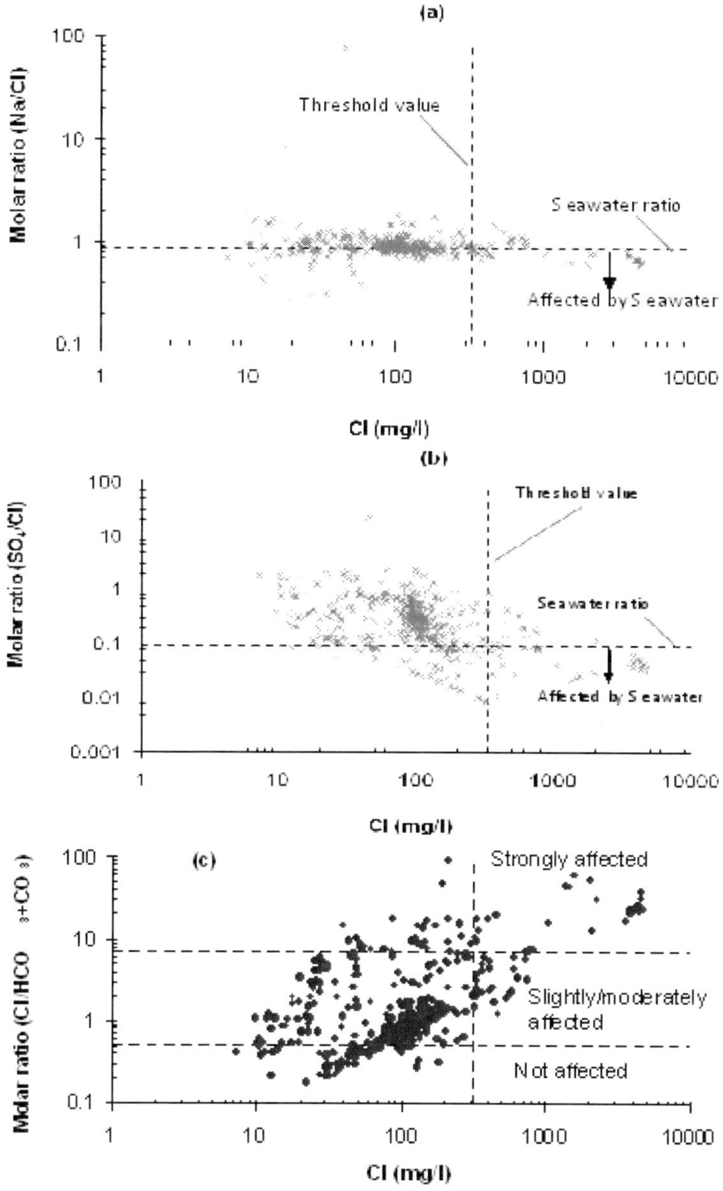

Figure 14. Molar ratios: (a) Na/Cl (b) SO_4/Cl; and (c) Cl/(HCO_3+CO_3) versus Cl concentrations in Cape Flats aquifer groundwater.

Shallow groundwater, much of which is affected by agriculture, may have undergone some degree of ET. This is due to the large input of irrigation water that undergoes transpiration by crops, and the resulting high water tables (locally < 1 m from the surface), which allow secondary ET (directly from the water table) to occur after recharge. This process is common in many semi-arid environments (Cartwright et al., 2006; Han et al., 2010; Currell et al., 2011).

Cation Exchange and Hydrogeochemical Evolution

Based on the preceding analyses, groundwater from the Cape Flats could be a mixture of recently recharged precipitation, which have undergone cation exchange reactions or mixed in relative proportions with seawater encroaching laterally from the coast. Major ion chemistry (e.g. molar ion ratios), are particularly useful in assessing sources of solutes and characterising hydrogeochemical evolution in aquifers (e.g. Edmunds et al. 1982; Herczeg and Edmunds 2000; Cartwright et al. 2004).

Cation exchange may also relate to large-scale disturbance and/or transient conditions in an aquifer (Currell et al., 2011); for example, high levels of pumping and/or mixing with irrigation water in recent decades may have mobilized Na that was otherwise relatively immobile in clay lenses. This would also add minor amounts of Na^+; however, the dominant source of sodium in groundwater is cation exchange with Ca^{2+} in the clay lenses of the deeper (Malmesbury) aquifer.

Exchange of ions is known to occur in different aquifer lithologies (e.g. Edmunds and Walton, 1983; Walraevens and Van Camp, 2005). As previously described, the quaternary deposits consist largely of aeolian sand, but minor fluvial to marine deposits also occur. The inter-relationships of the various formations are known from many boreholes and exposures within and outside the greater Cape Town area (Theron et al. 1992). Generally, cation exchange is favourable in relatively low ionic-strength waters (e.g. at [Na] < 0.1 M) where there is an abundance of negatively charged mineral surfaces, as these surfaces generally have greater affinity for divalent than monovalent cations (Stumm and Morgan, 1996). In the Cape Flats aquifer, the exchange likely occurs in lacustrine clay lenses interlayered with peat and argillaceous sand in places, which would contain abundant exchange sites. The shallow clayey and calcareous sands (widespread in the Cape Flats) also have a large potential primary Na-source, as the hyper-saline water (laterally from the coast) from which they were deposited had Na-rich chemistry (Adelana et al., 2010).

Cation exchange may additionally/alternatively occur within the loess that makes up the bulk of the aquifer matrix.

The occurrence of Na-Ca exchange in the aquifer has implications for the evolution of groundwater chemistry in the Cape Flats. Several other studies have shown mobilization of As and F in groundwater occurring due to changes in Na/Ca ratios in aquifer settings (Gomez et al., 2009; Currell et al., 2011). Despite the strong correlation between the concentrations of these major ions there is some variation in their ratios as well as in relation to other cations and molar ratios (Figure 15a-d). Molar Na/Cl ratios range from as high as 1.95 (Figure 15a), with many between 0.6 and 1.2 (essentially Cape Flats groundwater). The high Na/Cl ratios of the freshest groundwaters are probably controlled by water-rock interaction, for example, albite weathering by

$$2NaAlSi_3O_8 + 9H_2O + 2H_2CO_3 \rightarrow Al_2Si_2O_5(OH)_4 + 2Na^+ + 2HCO_3^- + 4H_4SiO_4 \quad (1)$$

(albite) (kaolinite)

Samples with Na/Cl ratios of 0.8–0.99 are probably due to evaporation of rainfall which has a Na/Cl ratio of 0.88±0.1 (Adelana et al., 2011). Samples with low Na/Cl ratios are probably the result of Na loss or exchange onto soils and clays. The high HCO_3/Cl ratios (up to 2.55, in the fresh groundwater, Figure 15b) may be due to the weathering of feldspars (as illustrated in equation 1), which may occur during the initial stages of water rock interaction.

Ca and Mg (Figure 15c-d) may have been derived from dissolution of silicate minerals (e.g. feldspar or biotite), and/or cation exchange of Na for Ca and Mg on clay minerals while dissolution of K-feldspar may be the major source of K^+. This is probably the case around Mitchells' Plain, very close to the coastline, where Ca is the dominant cation. There is the possibility that the groundwater rich in sodium ions (Na^+), as a result of seawater interaction, invades organic peat layers rich in calcium (Ca^{2+}). In this case, the sodium ions with a valency of (+1) replace the calcium with a valency of (+2). The reaction showing the cation exchange for the peaty layer is as follows:

$$2Na^+ + Ca - Peat_2 \rightarrow 2Na - Peat + Ca^{2+} \quad (2)$$

The effects of anthropogenic contamination in shallow groundwater (e.g. fertiliser application) may be reflected in the increasing concentrations of NO_3 in groundwater from 1983-1993 (Figure 16), which are mostly in excess of the drinking water standard of 10 mg/l (as NO_3-N). These are from wells around Ottery/Philippi/Mitchell's Plain area, where active farming activities are practised all-the-year round.

In contrast to salinity, groundwater flowing from the east has relatively low NO_3 levels. Agricultural activities are extensive in the northeast catchment of Macassar. This may be comparable to the intensive vegetable farming in the capture zone of the Cape Flats aquifer around Philippi/Ottery area, where >70% of vegetable production in the Western Cape are derived, and as such a good source of livelihood and income for this community dwellers. The current application rate of fertilizers has not been quantified but estimated as twofold higher than what it was in the 1980s (Adelana, 2011). There are no prescribed fertilizer application rate by law or restrictions on the quantity of fertilisers that can be applied to a given area of land, as it is in other parts of the world (CSUCE, 1994; USEPA, 2010; EA, 2013).

CONSTRAINTS AND MANAGEMENT IMPLICATIONS

From the samples analysed in this study, more than 85 per cent have Na and Cl concentrations below the WHO drinking water standard and essentially all the major, minor ions and trace elements are within the limits acceptable.

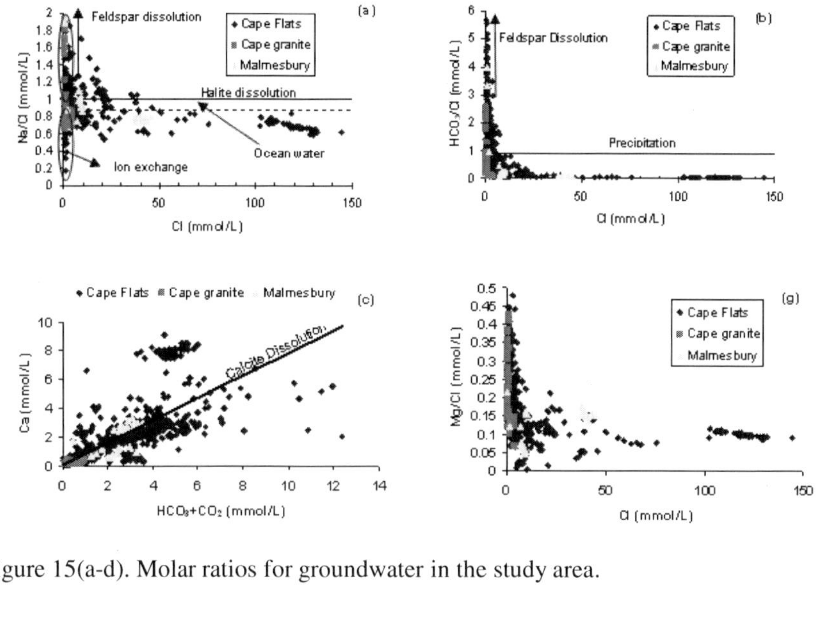

Figure 15(a-d). Molar ratios for groundwater in the study area.

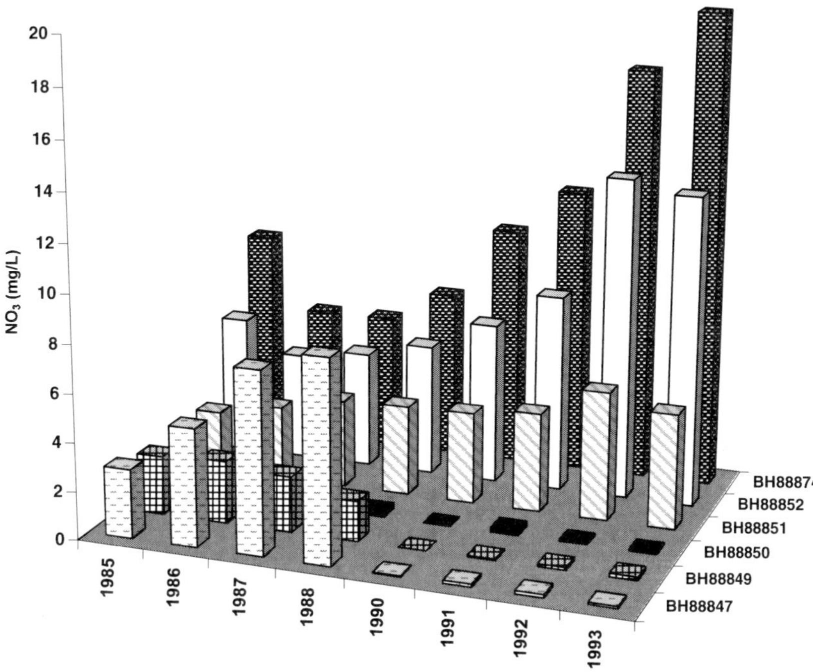

Figure 16. Variation in NO_3 concentrations in selected wells (in the Cape Flats).

This is an indication of overall good water quality and suitability for the purpose intended for the City of Cape Town. However, there are concerns about the level of nitrate concentrations in some areas, and high levels of chloride ions (close to seawater) in some boreholes within 10 km of the coast of False Bay.

Moreover, private (household) use of groundwater from the Cape Flats aquifer is widespread and increased from 2000-2007 (Colvin et al., 2007), especially since the increase in potable water tariffs in early 2000 (CCT, 2001). In response, the City of Cape Town has adopted an integrated approach to water management, which seeks a balance between water conservation and water demand management initiatives and conventional supply augmentation (Colvin et al., 2009).

Although the Department of Water Affairs (through the Cape Town regional office) in regulating groundwater development introduced strategies and management measures, the capacity-constraint within the department itself may limit the ability to continue groundwater monitoring. The cumulative effect of these could mean further deterioration of groundwater quality in the short and longer-term. Deterioration of groundwater quality if unchecked and ill-managed could limit the quantity of available (usable) groundwater in this area and may cause potential crisis for the local economy. Therefore, continuous monitoring is inevitable and detail investigation to determine the position of seawater wedge (e.g. through numerical modelling) with appropriate measures becomes necessary. The Cape Flats aquifer is vulnerable to many outside influences: the informal settlement of the Khayelitsha (poor sanitation) and the Philippi agricultural area (where fertilizer application takes place regularly) and numerous nodal sources of pollution (e.g. Waste Water Treatment Works, WWTW and the waste disposal sites). It may, therefore, be necessary to issue strong land use restrictions in order to prevent further contamination of groundwater. Public support/participation and education should be encouraged.

Finally, the dependence on groundwater is growing across the country. About 20% of water supplies in South Africa come from groundwater and over 60% of the population is dependent on it at some time of the year (DWAF, 2009). In recognition of this, the national strategy "Water for Growth and Development" through the Department of Water and Environmental Affairs" intends to grow groundwater supplies from nine to 10% by 2040 (DWAF, 2009). Therefore, adequate monitoring is required to keep the balance between water conservation and water demand.

SUMMARY AND CONCLUSION

The study examined physical and chemical characteristics of groundwater in the Cape Flats based on available historical data (1967-2001) and more recent updated sampling (2005-2007). Groundwater in the Cape Flats aquifer is predominantly of Na-Cl and Ca-Na-HCO_3-Cl type water. Whereas Na-Cl and Ca-HCO_3 types are dominant, various hydrogeochemical types were observed between the three aquifers of the area. Concentrations of major ions and ionic ratios were used to understand the evolution of groundwater chemistry within the Cape Flats aquifer system and its distinctions from other two aquifers (weathered Cape Granite suites and Malmesbury Shale).

Groundwater in the Cape Flats was categorised into four:

(1) 59.7% are dominantly influenced by the water-rock interaction;
(2) 2.4% of the groundwater samples are dominantly affected by the salinization process;
(3) 7.3% were affected by both water-rock interaction and salinization;

and

30.6% of the groundwater samples show negligible or low effects of salinization or water-rock interaction processes.

Most of the samples from Cape Flats aquifer have NaCl values below the WHO limits for drinking water (85.5% of samples below 250 mg/l Na and 79% are below 250 mg/l Cl^-), an indication of good water quality for the purpose of water supply augmentation intended for the City of Cape Town. However, there are concerns about some level of nitrate concentrations, resulting from agricultural practices and high levels of chloride ions in some boreholes indicating seawater encroachment from the coast and estuaries. Deterioration of groundwater quality from seawater influence could limit the amount of available sustainable groundwater in this area and may cause potential crisis for the local economy.

Reviewed by Prof. Tamiru Abiye, Professor of Hydrogeology, University of Witswaterstrand, Johannesburg, South Africa and Dr Jude Cobbing, former hydrogeologist at the British Geological Survey, a researcher and consultant in South Africa.

REFERENCES

Adelana, S. M. A. (2011). Groundwater resource evaluation and protection. Saarbrücken, LAMBERT Academic Publishing GmbH and Co., 296p.

Adelana, S. M. A.; Xu, Y. (2008). Impacts of land-use changes on a shallow coastal aquifer, South-Western Cape, South Africa. Proc. XXXVI Congress of the International Association of Hydrogeologists (IAH), Toyama, Japan, 28 October -2 November.

Adelana, S. M. A; Xu, Y; Adams, S. (2006). Identifying sources and mechanism of groundwater recharge in the Cape Flats, South Africa: Implications for sustainable resource management. Proc. XXXIV Congress of the International Association of Hydrogeologists (IAH), Beijing, China, 9-13 October 2006.

Adelana, S. M. A; Xu, Y; Vrbka, P. (2010). An integrated conceptual model for the development and management of the Cape Flats aquifer, South Africa. *Water SA*, 36 (4): 461-474.

Barber, C; Otto, C. J.; Bates, L. E.; Taylor, K. J. (1996). Evaluation of the relationship between land-use changes and groundwater quality in a water supply catchment, using GIS technology: The Gwelup wellfield, Western Australia. *Hydrogeology J.,* 4:6-19.

Cabrera, M. C; Custodio, E. (2004). Groundwater flow in a volcanic–sedimentary coastal aquifer: Telde area, Gran Canaria, Canary Islands, Spain. *Hydrogeology J.,* 10.1007/s10040-003-0316-y, Springer-Verlag.

Colvin, C. A.; Le Maitre, D.; Lotter, D. (2009). Climate Change and Water Resources. ScienceScope, CSIR.

Colvin, C.; Le Maitre, D.; Saayman, I.; Hughes, S. (2007). An introduction to Aquifer Dependent Ecosystems in South Africa. WRC TT.

Cartwright, I.; Weaver, T.; Fulton, S.; Nichol, C.; Reid, M.; Cheng, X. (2004). Hydrogeochemical and isotopic constraints on the origins of dryland salinity, Murray Basin, Victoria, Australia. *Appl. Geochem.,* 19:1233-1254.

City of Cape Town, CCT (2001) Water resources and water resource planning. Water Services Development Plan, December 2001, City of Cape Town.

Colorado State University Cooperative Extension, CSUCE. (1994). Best Management Practices for Nitrogen Fertilization (XCM-172). (1994, August). Retrieved February 19, 2014: http://www.ext.colostate.edu/PUBS/CROPS/pubcrop.html#soil

Currell, M. J.; Cartwright, I. (2011). Major-ion chemistry, $\delta^{13}C$ and $^{87}Sr/^{86}Sr$ as indicators of hydrochemical evolution and sources of salinity in groundwater in the Yuncheng Basin, China. *Hydrogeology J.*, 19:835-850.

Custodio, E. (2002). Coastal aquifers as important natural hydrogeological structures. In: Bocanegra E., Martinez D and Massone H. (eds.) *Groundwater and Human Development*, Mar de Plata, Argentina, pp.1905- 1918.

Djabri, L.; Laouar, R.; Hani, A.; Mania, J.; Mudry, J. (2003). The origin of water salinity on the Annaba coast (NE Algeria). Proceedings of symposium HSO2a held during IUGG2003 at Sapporo; IAHS Publ. No 280, pp. 229-35.

DWAF (2009). Water for Growth and Development Framework (version 7). Department of Water Affairs and Forestry, June 2009.

Edmunds, W. M.; Walton, N. R. G. (1983). The Lincolnshire Limestone - Hydrogeochemical evolution over a ten-year period. *J. Hydrol.,* 61:201–211.

Edmunds, W. M.; Bath, A. H.; Miles, D. L. (1982) Hydrochemical evolution of the East Midlands Triassic sandstone aquifer, England. *Geochim. Cosmochim. Acta.,* 46:2069-81.

Edmunds, W. M. (2012). Limits to the availability of groundwater in Africa. *Environ. Res. Lett.*, 7: 021003 (3pp) doi:10.1088/1748-9326/7/2/021003.

El Moujabber, M; Bou Samra, B.; Darwish, T.; Atallah, T. (2006). Comparison of different indicators for groundwater contamination by seawater intrusion on the Lebanese Coast. *Water Resources Management,* 20(2):161-180.

Environment Agency, EA (2013). Fertiliser Recommendations for Agricultural and Horticultural Crops, The Fertiliser Manual (RB209) on the Defra website http://www.defra.gov.uk/rb209

Gaaloul, N.; Pliakas, F.; Kallioras, A., Schuth, C., Marinos, P. (2012). Simulation of Seawater Intrusion in Coastal Aquifers: Forty Five-Years Exploitation in an Eastern Coast Aquifer in NE Tunisia. *The Open Hydrology Journal*, 6, (Suppl 1-M6) 31-44.

Gerber, A. (1976). An investigation into the hydraulic characteristics of the groundwater source in the Cape Flats. Unpublished M.Sc. thesis, University of the Orange Free State, Bloemfontein.

Gerber, A. (1981). A digital model of groundwater flow in the Cape Flats. CSIR Contract Report C WAT 46, Pretoria.

Giménez, E.; Morell, I. (1997). Hydrogeochemical analysis of salinization processes in the coastal aquifer of Oropesa (Castellón, Spain). *Environ. Geol.,* 29:118–131.

Green, E. D. (2006). Regional evaluation of saltwater intrusion of coastal aquifers in Caribbean Islands. Evaluation of the OAS-Funded Regional Project Report: Retrieved on May 20, 2014 from http://portal.oas.org/LinkClick.aspx?fileticket=9MUvoCFnOuM%3Dandtabid=1473

Güler, C., Thyne, G. D., McCray, J. E, Turner, A. K., 2002. Evaluation of graphical and multivariate statistical methods for classification of water chemistry data. *Hydrogeol. J.*, 18: 455-474.

Han, D. M.; Liang, X.; Currell, M. J.; Jin, M. G.; Zhong, W. J.; Liu, C. M.; Song, X. F. (2010). Environmental isotopic and hydrochemical characteristics of groundwater systems in Daying and Qicun geothermal fields, Xinzhou Basin, Shanxi, China. *Hydrol. Process,* 22:3157–3176.

Hendey, Q. B.; Dingle, R. V. (1983). Onshore sedimentary phosphate deposits in South Western Africa. Technical Report Joint Geological Survey/University of Cape Town Marine Geoscience Unit, 14, p27-40.

Henzen M (1973). The reclamation, storage and abstraction of purified sewage effluents in the Cape Peninsula (in Afrikaans). D.Sc. Thesis, University of the Orange Free State, Bloemfontein.

Herczeg, A. L.; Edmunds, W. M. (2000). Inorganic ions as tracers. In: Cook P, Herczeg A (eds) *Environmental tracers in subsurface hydrology*. Kluwer, Boston, pp 31–77.

Jeen, S. K.; Kim, J. M.; Ko, K. S.; Yum, B.; Chang, H. W. (2001). Hydrogeochemical characteristics of groundwater in a mid-western coastal aquifer system, Korea. *Geosci. J.,* 5:339-348.

Kan, A.; Xu, Y.; Usher, B. (2004). Hydrogeochemical Analysis Model (HAM) in excel: An overview of input, equations and use. Institute for Groundwater Studies, University of the Free State, Bloemfontein.

Lee, J. Y.; Song, S. H. (2006). Evaluation of groundwater quality in coastal areas: implications for sustainable agriculture. *Environ. Geol.,* 10.1007/ s00254-006-0560-2.

Malan, J. A. (1987). The Bredasdorp Group in the area between Gansbaai and Mossel Bay. *South African Journal of Science,* 83: 506-507.

Mercado, A. (1985). The use of hydrogeochemical patterns in carbonate sand and sandstone aquifers to identify intrusion and flushing of saline waters. *Ground Water,* 23:635-645.

Meyer, P. S. (2001). An explanation of the 1:500 000 hydrogeological map of Cape Town 3317. Department of Water Affairs and Forestry, 59pp.

Park, S. C.; Yun, S. T.; Chae, G. T.; Yoo, I. S.; Shin, K. S.; Heo, C. H.; Lee, S. K. (2005). Regional hydrochemical study on salinization of coastal aquifers, western coastal area of South Korea. *J. Hydrol.,* 313:182-194.

Park, S. C.; Yun, S. T.; Chae, G. T.; Lee, S. K. (2002). Hydrochemistry of shallow groundwaters in western coastal area of Korea: a study on seawater mixing in coastal aquifers. *J. KoSSGE.,* 7:63-77.

Petalas, C. P.; Diamantis, I. B. (1999). Origin and distribution of saline groundwaters in the upper Miocene aquifer system, coastal Rhodope area, northeastern Greece. *Hydrogeol. J.,* 7(3):305–316.

Pulido-Bosch, A.; Tahiri, A.; Vallejos, A. (1999). Hydrogeochemical characteristics of processes in the Temara Aquifer in Northwestern Morocco. *Water Air Soil Pollut.,* 114: 323-337.

Richter, B. C.; Kreitler, C. W.; Bledsoe, B. E. (1993). Geochemical techniques for identifying sources of ground-water salinization. CRC, New York.

Sadeg, S. A.; Karahanoglu, N. (2008). Numerical assessment of seawater intrusion in the Tripoli region, Libya. *Environ. Geol.,* 40: 1151-68.

Sanchez, M. F.; Pulido-Bosch, A.; Sanchez, L.; Vallejos-Izquierdo, A. (2002). Identification of the origin of salinization in groundwater using minor ions (Lower Andarax, Southeast Spain). *Sci. Total Environ.,* 297:43-58.

Sinclair, A. J. (1974). Selection of thresholds in geochemical data using probability graphs. *J. Geochem. Explor.,* 3:129-149.

Stuyfzand, P. J. (1999). Patterns in groundwater chemistry resulting from groundwater flow. *Hydrogeol. J.,* 7(1):15-27.

Theron, J. N. (1974). Geological sheet 3418BA, Strandfontein (1:50 000): Geological Survey, South Africa (Open file).

Theron, J. N., Gresse, P. G.; Siegfried, H. P.; Rogers, J. (1992). The geology of the Cape Town area. Explanation on Sheet 3318, Geological Survey, South Africa, 140p.

USEPA (2010). Managing Agricultural Fertilizer Application to Prevent Contamination of Drinking Water. Source Water Protection Practices Bulletin, USEPA East (EPA East), August 2010, Washington, DC 20004.

Unites States Geological Survey, USGS. (2007). Freshwater-Saltwater Interactions along the Atlantic Coast. Retrieved on February 6, 2008 from http://water.usgs.gov/ogw/gwrp/saltwater/.

United States Geological Survey, USGS. (2005). Saltwater Intrusion in Los Angeles Area Coastal Aquifers- The Marine Connection. Retrieved on February 6, 2008 from http://pubs.usgs.gov/fs/2002/fs030-02/

United States Geological Survey, USGS. (2005). Southern California Salt-Water Intrusion. Retrieved on February 6, 2008 from http://walrus.wr.usgs.gov/cabrillo/casa/

Vandoolaeghe, M. A. C. (1989). The Cape Flats groundwater development pilot abstraction scheme. Technical Report No. GH3655, Directorate Geohydrology, DWAF, Cape Town.

Walraevens, K.; Van Camp, M. (2005). Advances in understanding natural groundwater quality controls in coastal aquifers. Groundwater and saline intrusion, selected papers from the 18th Salt Water Intrusion Meeting, Cartagena 2004, Spain, pp 449–463.

In: Groundwater
Editor: Segun Michael Adelana
ISBN: 978-1-63321-759-1

Chapter 4

OVERDEVELOPMENT OF GROUNDWATER AND EFFECT ON ITS MOVEMENT AND QUALITY IN AN AGRICULTURE DOMINATED AREA IN NORTH-WESTERN INDIA

S. K. Goyal[1,*], A. K. Narula[1] and B. S. Chaudhary[2]
[1]Department of Physics, RKSD College, Kaithal, Haryana, India
[2]Department of Geophysics, Kurukshetra University, Kurukshetra, Haryana, India

ABSTRACT

Agriculture from irrigated land has contributed 13.7% of India's gross domestic product (GDP) during 2012-13. Kaithal district (north-western India) represents an agriculture dominated area. The groundwater development in the district has taken place without proper understanding of its occurrence in time and space. Therefore, risks related to groundwater levels are a cause of serious concern. The present study analysed the effect of over exploitation on sustainability of the groundwater resource. Depth to water level and quality data from 79 observation wells from 1974 to 2009 revealed that excessive pumping has resulted in lowering of groundwater levels in most parts, particularly in fresh quality zone. Average depth to water level below ground level (bgl)

* Corresponding author: S. K. Goyal. Department of Physics, RKSD College, Kaithal, Haryana, India-136027. E-mail: sanjayktl@gmail.com.

in the district was found to recede from 6.21 m in 1974 to 19.16 m in 2009. The net decline was found to be 11 m from 1999 to 2009 against a decline of 1.5 m from 1974 to 1999. A comparison of the post-monsoon (October) groundwater levels with the respective pre-monsoon (June) levels showed that from 1999 onwards, the water table depicted no net replenishment during monsoon. The spatial maps of 1987, 1997 and 2007 showing the elevations of groundwater levels above mean sea level indicated a shift in the groundwater flow from north-east to south-west (in 1987) to that from south to north-west (in 2007). The changes in groundwater quality during this period were also found to corroborate with these observations. The study visualizes a potential threat to the quantity and quality of groundwater which may cause serious socio-economic consequences.

Thus the resource requires judicious management and regulations to arrest the declining trend of groundwater levels.

Keywords: Groundwater, Depth to water level, Quality, GIS, Agriculture, Irrigation

INTRODUCTION

Worldwide, agriculture accounts for 70% of the total water consumption, compared to 20% for industry and 10% for domestic use (UN, 2006). The advent and rapid spread of energized pumping technologies have enabled speedy groundwater development which has led to the emergence of land use and cropping systems dependent on its reliability. This has helped in bringing green revolution in India, particularly in the north-western region. Agriculture from irrigated land has contributed 13.7% of India's gross domestic product (GDP) during 2012-13 (CSO, 2013). Therefore, risks related to groundwater levels are a cause of serious concern. Groundwater is threatened by over-exploitation as its development has taken place without proper understanding of its occurrence in time and space (CGWB, 2006); with cases of declining water tables reported widely, particularly in densely populated areas (Narain, 1998). An increase in demand has led to higher withdrawal of groundwater. Abstract ion exceeding natural replenishment has generated stress in aquifers causing depletion of the water table.

The factors contributing to unsustainable use of groundwater resources and its degradation in different forms are varied and spatially specific. Excessive withdrawal to meet the needs of increasingly intensive cropping is leading to lowering of water tables in several areas.

This is also causing the rise in water table in areas where its use is restricted due to quality considerations, leading to spread of salinity problems (Prabhakaran et al., 2009). Moreover, climate variability also makes a direct impact on groundwater by way of change in recharge and aquifer recovery. The groundwater levels may take months or years to replenish once pumped for irrigation or other uses. The lowering of groundwater levels has resulted in reduction in individual well yield, growth in well population, failure of bore-wells, drying up of dug-wells, increase in power consumption, changes in direction and velocity of groundwater flow and ecological damage (Imtiyaz and Rao, 2008).

Further, the environmental effects may also cause reduction of porosity, deterioration of water quality, loss of vegetation, land subsidence, inland ingress of saline water in coastal regions etc. There is a strong relationship between development and management of groundwater resources. The optimal use of this resource for a range of diversified and increasing demands requires a good understanding of the issues involved. For example, a study carried out in a small settlement in Karnataka state showed that out of 84 bore-wells in use during 1980-1992, only 27 were found operational by 1993 (Ballukraya, 1997). Analysis of water levels revealed that groundwater resources in Dhund basin are dwindling (Sinha and Singh, 2006). An assessment of total water resources available at block level in Ludhiana district during *Rabi* and *Kharif* seasons showed a continuous declining trend during 1994-95 to 2004-05 (Kaur et al., 2009). Stage of groundwater development and causes of water table decline have been studied to develop models for predicting water table depth in Ramganga - Bahgul inter basin of Uttar Pradesh (Sarkar et al., 2007). A study on water table trends from 1984 to 1998 through 53 permanent hydrograph stations in West Bengal found a positive correlation with yearly rainfall trends (Biswas, 2003). Dynamics of groundwater flow under different hydrogeological conditions in time and space have been studied in relation to steep decline of water level and sharp deterioration in its quality (Jain, 2009).

Gravity of the problem has increased from a situation of over-exploitation to that of severe environmental degradation and water security. Therefore, the present study highlights that demand driven exploitation without understanding the area specific problems and lack of regulatory measures has led to the present crisis and damage to groundwater systems. The study was carried out in Kaithal district with the objective to analyse spatio-temporal changes in groundwater depth and quality to assess the effect of overdevelopment on sustainability. The analysis was done in GIS environment using Integrated Land and Water Information System (ILWIS) software.

The Study Area

Kaithal district is located in north-western part of the state between latitudes 29° 32’ to 30° 12’ north and longitudes 76° 08’ to 76° 45’ east (Figure 1).

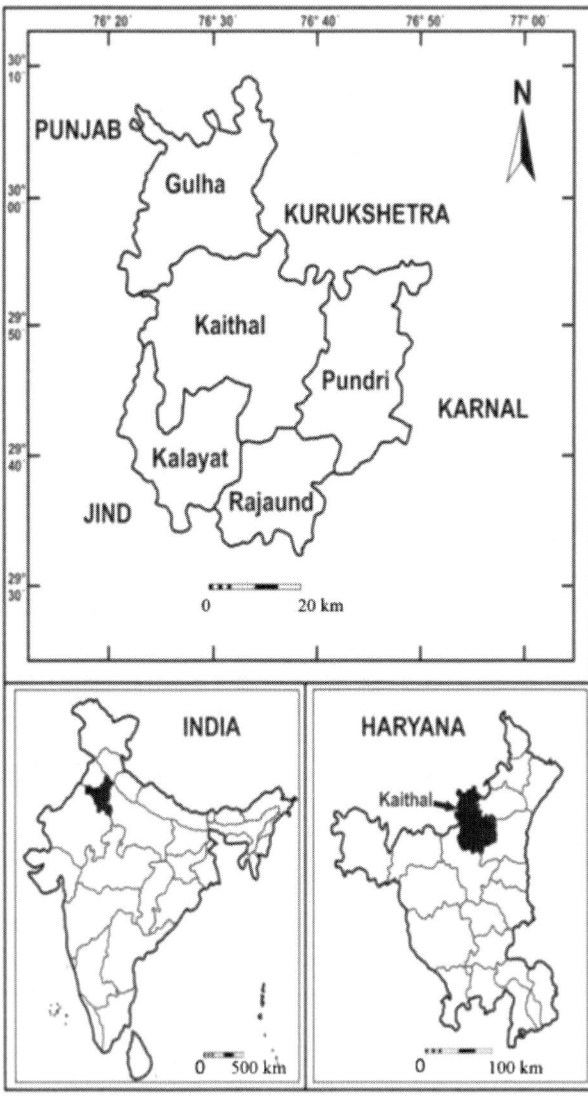

Figure 1. The Study area.

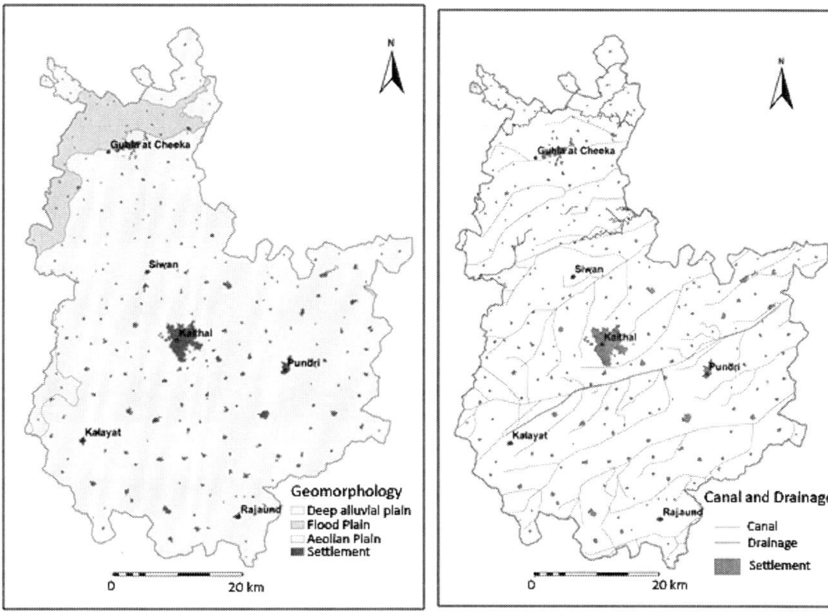

a b

Adapted from: HARSAC, Hisar.

Figure 2. (a) Geomorphology; and (b) Natural drainage and distribution of canals in Kaithal district.

It covers 2317 km² of area and comprises of 276 revenue villages distributed in five administrative blocks namely Gulha, Kaithal, Pundri, Kalayat and Rajaund. The district had a population of 10, 72, 861 (Census of India, 2011).

On the basis of genesis and physical characteristics, the district has been classified into three units: flood plain, alluvial plain and aeolian plain (HARSAC, 2004). A flat to gently sloping strip along the river Ghaggar, passing through northern part of the district represents the flood plain (Figure 2a). The zone covers about 183 km^2 area and comprises of fluvial deposits like sand, silt and clay sized particles. About 90% of the total area is covered by alluvial plain. A small pocket of undulating aeolian plain is also found in Kalayat block.

The area is a part of the vast Indo-Gangetic alluvial plain of pleistocene to recent series in quaternary system. The topography is flat, devoid of any prominent feature. The general elevation varies from 217 m to 252 m above mean sea level, with a gentle slope from north-east to south-west. The district can be divided in two units namely upland and lowland plains.

The upland plain is spread along the north-eastern boundary of the district and includes Gulha, Kaithal and Pundri blocks. Kalayat and Rajaund blocks constitute the low lying plain. The thickness of the alluvium varies from 250 m to 805 m (Singh, 1992).

The lithology comprises of fine to medium sand with alternate clay beds. The newer alluvium consists of windblown sand, fine sand and silt while the older alluvium is constituted by fluvial deposits, clay, silt, sand and gravel. The district is characterised by the presence of thick sandy aquifer zones intervened by thin layers of clay.

These zones can be identified as a single unified multi-aquifer system. The aquifers are either in the form of isolated lenses of sand embedded in clay beds or well-connected granular zones (Sharma, 2008). Groundwater occurs under water table conditions in saturation zone.

The movement of groundwater is from north-east to south-west, following natural slope of the surface (Singh, 1992). The depth of shallow tubewells generally ranges from 20-60 m, yielding a discharge rate of more than 10 litres per second (lps) (Table 1). The position of pumps generally ranges up to 20 m in southern blocks and above 20 m in northern blocks.

These are operated with the help of prime-movers of 5-10 horse power (HP) in southern blocks and above 10 HP in northern blocks. Major sources of groundwater recharge are: inflow, rainfall, seepage from canals, return seepage through irrigation and percolation from surface-water bodies.

The soils in the district belong to arid brown group. It is calcareous in nature with a layer of calcium carbonate nodules occurring at a depth of 1 to 1.5 m. The soil texture varies from sandy loam to loam on the surface and from loam to clay loam or silty clay loam in the subsurface in upland areas.

On the other hand, in low lying areas, the soils are poorly drained and soil texture varies from loam to clay loam or silty clay.

The land use in the district is dominated by agriculture. The agricultural practices have been influenced by innovations in technology. More than 88% of total district area is under cultivation. Rice and wheat are major crops. The area under rice and wheat in 2010-11 was 158.7 and 177.6 thousand hectares accounting for 7.2 and 12.6 % of the total production of the state respectively (Statistical Abstract, 2013). More than 80% of the area is double cropped.

However, in some parts of Kalayat and Rajaund blocks, only Rabi crop is grown. Due to decreasing landholding and increasing pressure on available agricultural land, practice of growing additional short duration crop (known as Saathi or summer paddy) is prevalent in Gulha, Kaithal and Pundri blocks.

Table 1. Block wise area showing discharge of tubewells in different ranges

Block	Total area (km^2)	Area under different ranges of discharge (km^2)		
		8-10 lps	10-12 lps	>12 lps
Gulha	593	-	-	593
Kaithal	635	18	91	526
Kalayat	323	248	72	3
Pundri	448	15	134	299
Rajaund	285	240	36	10

Source: Groundwater Cell, Department of Agriculture, Government of Haryana.

Table 2. Block wise utilization of groundwater in Kaithal district (ha-m)

Block	Irrigation	Industrial use	Domestic use
Gulha	25570	720	47
Kaithal	22674	1035	40
Kalayat	8141	0	5
Pundri	26619	585	20
Rajaund	15151	45	0

Source: Groundwater Cell, Department of Agriculture, Kurukshetra.

Table 3. Number of diesel and electric units in Kaithal district

Tubewell sets	1991	2002	2007	2012
Diesel units	15713	23194	23819	19365
Electric units	20381	27525	33301	42542
Total	36094	50719	57120	61907

Statistical abstract, 2013.

The climate of Kaithal district is sub-tropical, semi arid, continental and monsoon type. It is characterized by hot summers, cool winters and a short rainy season. Winter season starts towards the latter half of November and extends till the middle of March. The summer continues from April till the end of June. During summers, the maximum temperature reaches up to 45 °C. The rainy season remains active from July to September. The climate is suitable for crops like wheat, rice and sugarcane. The average annual rainfall is about 433 mm. Rainfall is received mainly from south-west monsoon. July and August are the wettest months, contributing more than 70% of annual rainfall.

Due to dominant cultivation of water intensive crops, particularly paddy, there has been a tremendous demand for irrigation water. A network of distributaries and minors provides surface-water in the district. However, the canal network is limited and concentrated largely in southern parts as compared to northern parts (Figure 2b). Moreover, the supply of surface water is not sufficient due to natural physiography, high population density, limitations of the canal based supply, and intense agricultural activity.

Consequently, there is a tremendous dependence on the groundwater resource, mainly for irrigation (Table 2). The groundwater development (Table 3) in the district has posed a potential threat to the quality and availability of groundwater.

METHODS AND MATERIALS

The long term pre- and post-monsoon observations of depth to water level below ground level (bgl) along with other collatoral data have been used to study the groundwater dynamics. Further, the physico-chemical analysis data of groundwater samples was used to compute the parameters that determine the suitability of groundwater for domestic and agricultural use.

The measurements of depth to water level bgl from selected observation wells were carried out in June (pre-monsoon) and October (post-monsoon). The archive data from 1974 to 2007 was obtained from Groundwater cell, Government of Haryana. The data was processed further keeping in view the consistency of readings and location of observation points. 79 out of 108 observation wells were selected for the final analysis. Locations of these wells were random but covered the entire district. Elevations of groundwater levels (reduced levels) from mean sea level (msl) were also computed by subtracting the depth to water level bgl from surface elevations. The physico-chemical analysis data of groundwater samples from 1987 to 2007 was also obtained.

Boundary map of Kaithal district was digitized in GIS environment with the help of Survey of India topographical maps 53 B/4, B/8, C/1, C/2, C/5, C/6, C/9 and C/10 at 1:50000 scale (Survey of India). Digitization was carried out to obtain various vector layers namely district boundary, block boundary and well point location. The point data pertaining to elevations of observation wells from mean sea level, depth to water levels and groundwater quality were organized in the form of attribute tables and attached with the point map representing the locations of observation wells. To estimate the values of these parameters at unsampled locations, interpolations were carried out using the

Moving Average method with ILWIS 3.6 software. The spatial distribution maps thus prepared were further classified into various zones by using *Slicing* operation. For classifying the depth to water level maps, various depth zones were defined as shallow, moderate, deep and very deep for depth to water level in the ranges 0-3 m, 3-10 m, 10-30 m and 30-45 m respectively.

To classify groundwater quality maps into suitability zones for domestic and agricultural uses, standard criteria were followed. Areas under different classified zones in the maps were determined by generating histograms for the *sliced* maps. *Cross* and *Difference operations* in ILWIS were used to prepare various *cross* and *difference maps* for analysis.

Depth to water level bgl data obtained from the observation wells were converted into water level elevations above msl by subtracting from surface elevations at the respective points. The lateral variation map for groundwater elevations was prepared by point interpolations.

Further, these maps were classified into different elevation zones at 5 m intervals. The tentative directions of groundwater flow were drawn from higher elevation zone to lower elevation zone (Biswas, 2003).

RESULTS AND DISCUSSIONS

Long Term Spatio-Temporal Variability of Groundwater Levels

The spatial distribution of groundwater depth was mapped and classified into various ranges at intervals of five years from 1987 to 2007 (Figure 3). The area under different depth zones was determined (Table 4). The results depicted a significant temporal change in areas under different depth zones.

During the period under study, a decrease in areas under *normal* and *moderate* zones was observed.

Table 4. Area under different depth ranges in Kaithal district (1987-2007)

Year	0-3 m		3-10 m		10-20 m		20-30 m		30-40 m	
	(km^2)	%	(km^2)	%	(km^2)	%	(km^2)	%	(km^2)	%
1987	37.5	1.6	1735	76.0	511	22.4	0	0	0	0
1992	0	0	1242	54.4	1042	45.6	0	0	0	0
1997	10	0.4	1278	56.0	983	43.0	13	0	0	0
2002	0	0	909	40.0	1138	49.8	234	10.3	1	0
2007	0	0	657	28.8	808	35.4	627	27.5	191	8.4

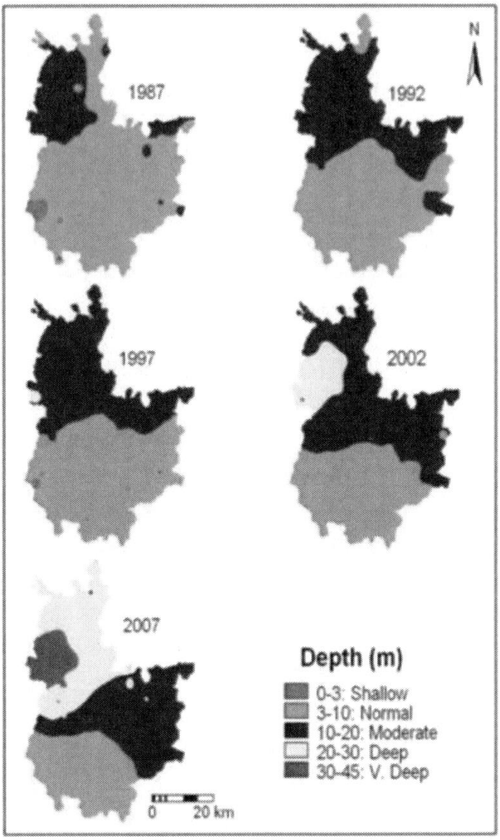

Figure 3. Spatial variation of groundwater depth in Kaithal district.

However, areas under *deep* and *very deep* zones were found to increase. It was also revealed that the decline in groundwater levels was not spatially uniform. A regular decline was observed in more than 70% of the area (maximum in north-western region). However, in the remaining part (southern region) groundwater levels were found fluctuating in a rather narrow range of 4-9 m.

Decennial Depletion Rates

The difference maps (1997-1987) and (2007-1997) were prepared to depict the changes in spatial distribution of depth to water levels during corresponding periods (Figure 4).

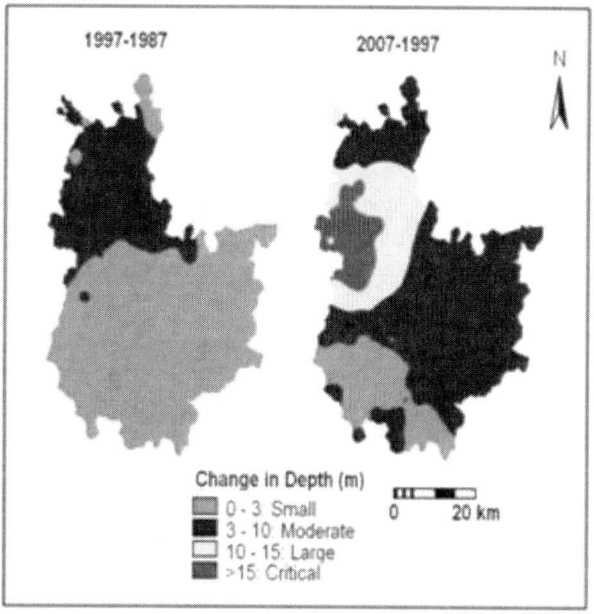

Figure 4. Change in depth to water levels during 1987-97 and 1997-2007.

During 1987-1997, a *moderate* decline of 3-10 m in groundwater levels was observed in northern part against a *small* decline of 0-3 m in the remaining part of the district. In succeeding decade (1997-2007), the decline was found to be *moderate* to *large* in major part of the district. *Critical* depletion by more than 15 m was also observed in a pocket in north–western part. A comparative analysis of depletion in the study area during two successive decades showed that the problem has aggravated in the last decade. This can be attributed to a decreased rainfall accompanied by a spurt in use of submersible systems.

The general depletion of water table may be due to over-exploitation to meet the huge irrigation requirements of intensive cultivation of paddy. The area under cultivation of rice has increased from 1090 km^2 in 1989-90 to 1510 km^2 in 2006-07 (Statistical Abstract, 2013). A comparatively high depletion in northern and north-western areas may be attributed to spatial disparity in physiography, soil, drainage, quality of groundwater, canal network and groundwater development. These conditions suggest greater abstraction and smaller recharge in northern areas as compared to southern parts. Moreover, the alarming depletion observed in north-western part could be attributed to cultivation of an additional short duration crop (summer paddy) in certain areas (particularly Gulha block) due to high evapo-transpiration losses, lack of rainfall and complete dependence of the crop on groundwater (Yadav et al., 2008). The observed decline has been so fast that if this trend continues unchecked, consequences may include collapse of agricultural output and severe shortage of potable water.

Block-Wise Analysis of Pre- and Post-Monsoon Levels

Figures 5-9 depict the block-wise variation of average pre- and post-monsoon depth to water level bgl from 1987 to 2007. The pre- and post-monsoon depth to water level in Gulha block (Figure 5) was found to be almost equal during 1987 to 1997. During the next decade, the post-monsoon groundwater levels were found to be lower than the pre-monsoon levels. Moreover, this trend appeared to be increasing from 1998 to 2007. However, in other blocks, the corresponding post-monsoon levels were observed to be higher. This showed that from 1987 to 1997, the monsoonal recharge was able to completely replenish the depletion occurred throughout the year in Kalayat, Rajaund and Pundri blocks and to some extent in Kaithal block. In Gulha block, the net draft has been more than net recharge.

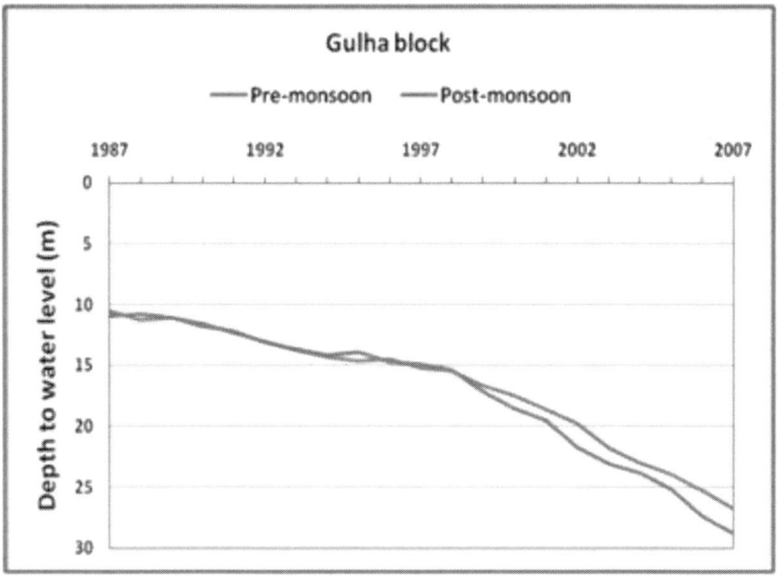

Figure 5. Variation of average depth to water levels in Gulha block.

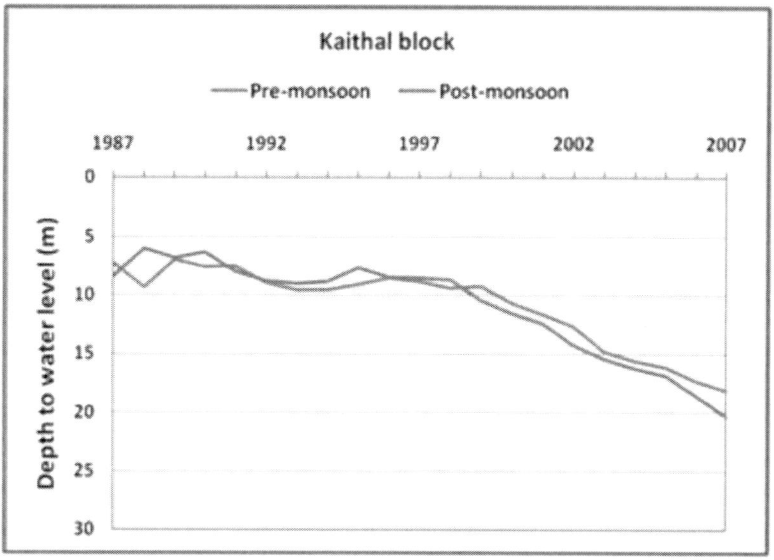

Figure 6. Variation of average depth to water levels in Kaithal block.

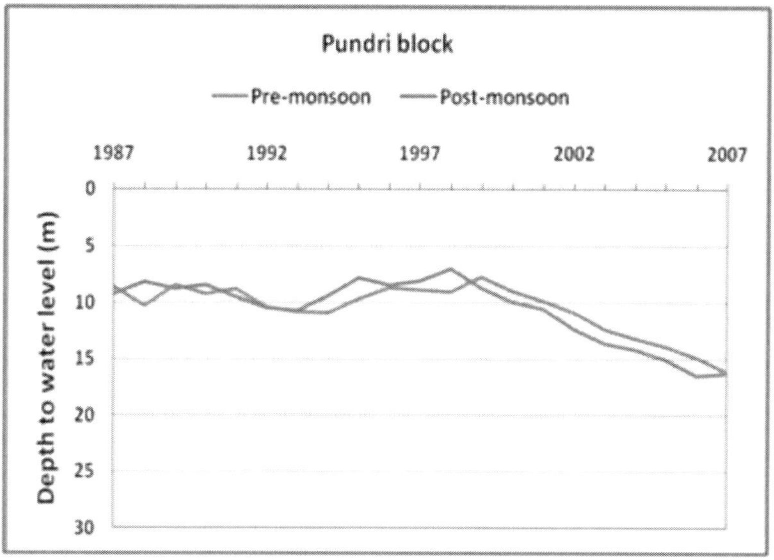

Figure 7. Variation of average depth to water levels in Pundri block.

Also, after 1999, the post- monsoon levels were found below the corresponding pre-monsoon levels in Kaithal and Pundri blocks. Increased pumping capacity of tube wells might have contributed to this trend.

Low lying Kalayat and Rajaund blocks demonstrated almost stable water table. Figure 10 shows the district average of pre- and post-monsoon trends of depth to water level bgl. It revealed a pattern of general decline of groundwater levels.

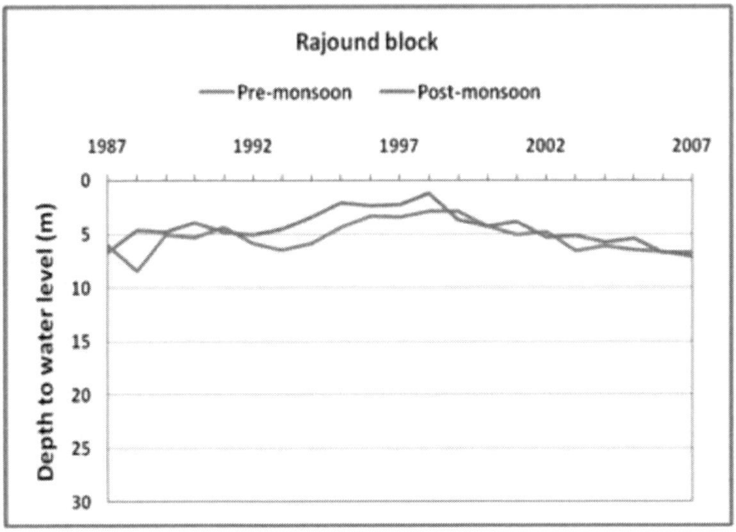

Figure 8. Variation of average depth to water levels in Rajaund block.

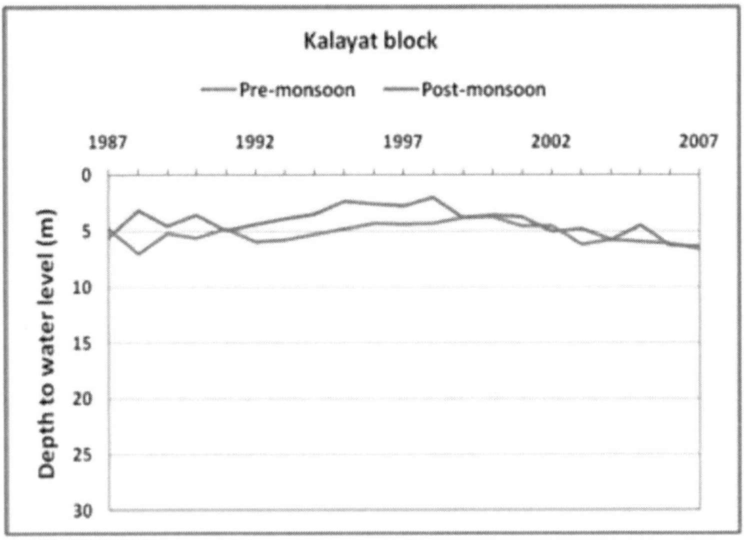

Figure 9. Variation of average depth to water levels in Kalayat block.

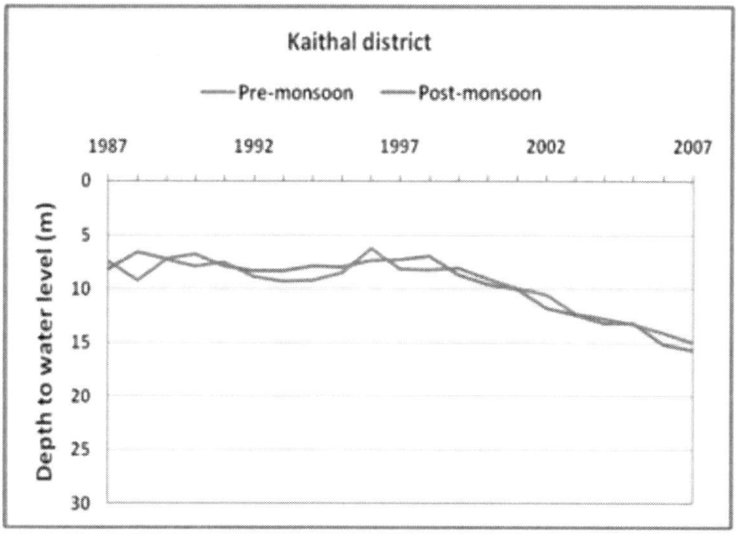

Figure 10. Variation of average depth to water levels in Kaithal district.

Comparatively higher pre-monsoon levels in 1989 and 1996 may be attributed to heavy rainfall in 1988 and floods in 1995. From 1987 to 1998, the average depth appeared to be around 7 m. It was found to decline from 7 m in 1987 to 15 m in 2007 (0.4 m per year), in agreement with Rodell et al., 2009.

Further, a significant increase in area as well as extent of net groundwater decline was observed from 1987 to 2007. This showed that over a period of two decades, the over-exploitation has increased to an extent that natural recharge during the monsoon season was not sufficient to replenish the levels.

Thus, there was a scope for artificial recharge in such parts. The cropping pattern, increase in tubewell density, position of pumps in shallow tubewells, area under submersible pumps, power capacity of prime movers and net average discharge indicated the dependence on groundwater in different blocks. The analysis of agriculture and groundwater development data revealed that the magnitude of over-exploitation was maximum in Gulha block, followed by Kaithal and Pundri blocks.

Changes in Directions of Groundwater Movement

The spatial map showing the elevations of groundwater levels above msl in the district were compared at ten year intervals from 1987 to 2007. In 1987, the water table elevation showed a general downward trend from north-east to south-west except a depression zone in northwest (Figure 11). The elevations appeared to agree with physiography of the area. The flow of groundwater was along the surface slope. However, a perceptible change in water table elevations and groundwater flow directions was observed in 1997.

These changes were found more prominent and conspicuous in 2007. It was thus inferred that the groundwater elevations have changed significantly. This has, probably led to a shift in the general flow of groundwater from north-east to south-west in 1987 to that of towards north-west in 2007.

Suitability of Groundwater Quality for Domestic Use

The *crossmap* obtained from lateral variation maps of pH, TDS and TH depicted the groundwater suitability for domestic use (Figure 12). The study area was found divided into different zones of suitability.

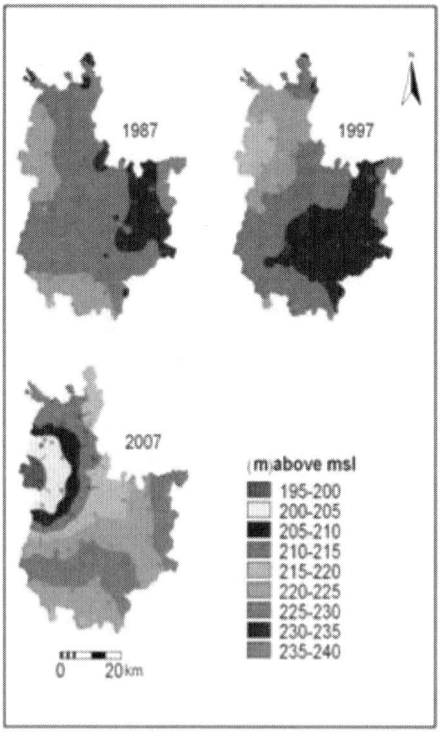

Figure 11. Maps showing tentative groundwater flow directions in Kaithal district.

Figure 12. Suitability of groundwater for domestic use in Kaithal district.

Table 5. Area underground water suitability zones for domestic use in Kaithal district

Quality class	Area (km^2)	Area (%)
Good	2018	88
Moderate	230	10
Doubtful	35	2

About 88% area was found to be under *good* class of quality, whereas in rest 12% of area falling under Kalayat and Rajaund blocks was under moderate to *doubtful* quality due to TDS and pH exceeding maximum permissible limits (Table 5). In general, groundwater in Kaithal district was found to be hard, alkaline and saline in nature. No health-based concern was percieved in groundwater use for drinking purpose. However, it was of low public acceptibility in many villages of Rajaund and Kalayat blocks.

The quality problems faced by these villages can be correlated with raised water table and water logging conditions prevailing at certain places due to physiographic depression (Goyal et al., 2006).

Significant increase in TDS and negligible changes in pH and TH were aslo observed. Moreover, the monsoonal recharge has been found to be leading to slight improvement in quality.

Suitability of Groundwater Quality for Agricultural Use

The resultant composite map depicts the study area into different suitability zones for agricultural use (Figure 13). For irrigation, the quality was found good in 34%, *moderate* in 27%, *doubtful* in 30% and *non-suitable* in 8% of the total district area (Table 6). All the water quality parameters showed a perceptible moderation during the post-monsoon observations. This dilution may be attributed to replenishment of the groundwater by rainfall during the monsoon season. Moreover, the relative improvement in groundwater quality during post- monsoon observations indicated a clean recharge without pollution from external sources.

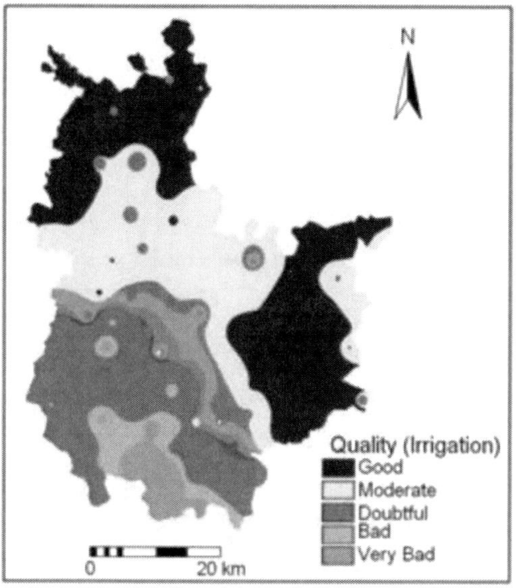

Figure 13. Suitability of groundwater for agricultural use in Kaithal district.

Table 6. Area under different zones of groundwater suitability for agricultural use in Kaithal district

Quality class	Area (km^2)	% of total district area
Good	773	34
Moderate	620	27
Doubtful	680	30
Bad	98	4
Very bad	94	4

Changes in Groundwater Flow Directions and Deterioration of Quality

Under natural conditions, the boundary between the freshwater and saltwater tends to be relatively stable, but pumping can cause saltwater to migrate inland and upward, resulting in contamination of the water supply (Gleick, 1996). The status of groundwater quality during 1987 and 2007 has been depicted by EC lateral variation maps pertaining to the corresponding years (Figure 14).

It was found that during 1987, northern and eastern part of the district had EC in *good quality* range (250-750 μS/cm). The central part had *moderate*, whereas southern part had *poor* to *very poor* quality. A significant deterioration was found in the *good* quality areas.

However, the quality in the southern areas showed negligible changes during 1987-2007. The difference map obtained by subtracting 1987 EC map from the corresponding 2007 map quantified these changes (Figure 15 a).

A general increase in EC was found in entire district except in southern areas. However, a higher increase, ranging between 1000-2000 μS/cm was found in northern, north-western and south-western parts.

The general deterioration of groundwater quality can be explained by the effect of the heavy application of fertilizers and their subsequent leaching under poor drainage conditions (Chae et al., 2004).

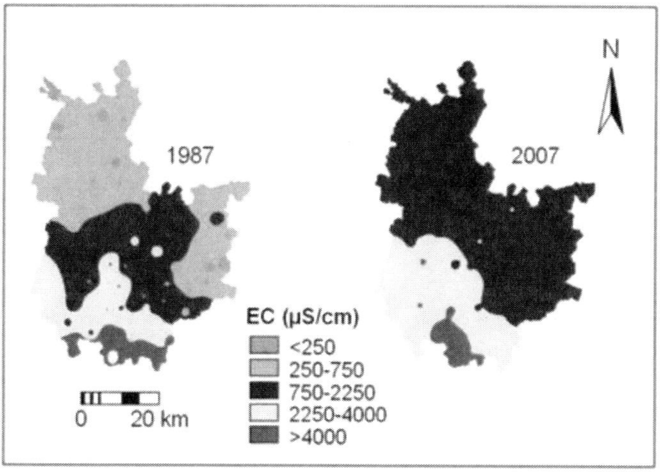

Figure 14. Lateral variations of EC in Kaithal district in 1987 and 2007.

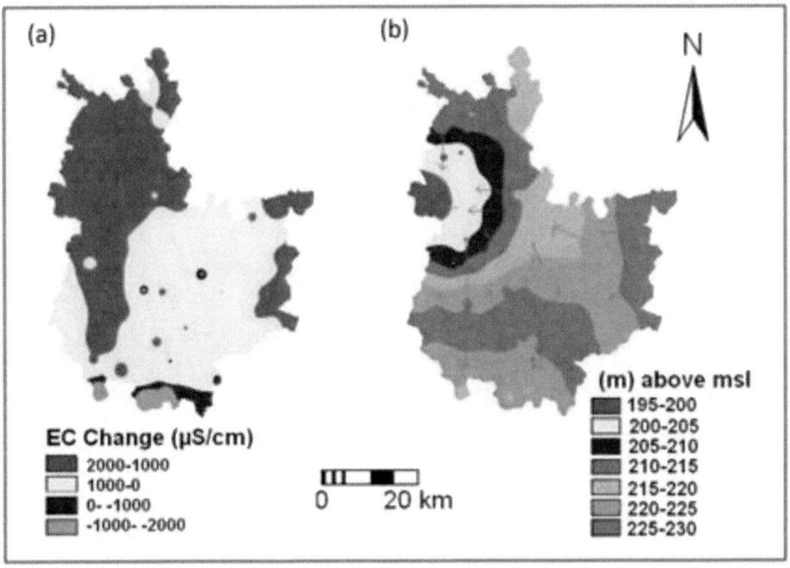

Figure 15. (a) Change in EC from 1987 to 2007 and (b) direction of groundwater flow in Kaithal district.

The relatively large deterioration observed in north-western and south-western parts can be associated with changes in groundwater flow directions led by altered groundwater elevations due to over-exploitation (Rotaru et al., 2004). The spatial map in Figure 15b showed the elevations of groundwater levels above msl along with tentative directions of groundwater flow during the year 2007. The general flow of groundwater in 2007 was found to shift towards north-west direction from its natural movement along south-west. This has probably caused the intrusion of saline water from southern areas to other parts resulting in deterioration of quality in fresh-water zone.

CONCLUSION

The present study has revealed problems of lowering of groundwater levels and deterioration of its quality due to over-exploitation, indicating a potential threat to sustainability. While the decline in groundwater levels was regular and significant in two-third area that included Gulha, Kaithal and Pundri blocks. The average rate of depletion was observed to be highest in Gulha block.

It was revealed that about 10% of the district area falling under Gulha block had undergone a decline by more than 1 m per year. Therefore, this part was identified as critical zone.

The depth to water level and rate of depletion were found to be moderate in the north-eastern and central part which includes entire Pundri block and a part of Kaithal block. The spatio-temporal analysis of pre- and post- monsoon groundwater levels revealed a significant decrease in amount of net groundwater recharge during monsoon particularly from 1998 to 2007. The groundwater levels in Kalayat and Rajaund blocks showed negligible decline.

The study indicated creation of a depression zone in north-western part. Therefore, a water table gradient towards north-western direction was observed. This may have led the groundwater to flow from south and south-east to this depression zone. This observed shift in direction of groundwater movement points to serious environmental degradation.

The quality of groundwater was found to be good to moderate in Gulha, Pundri and Kaithal blocks, whereas doubtful to non-suitable in Rajaund and Kalayat blocks. A part of Rajaund block was identified as critical having high value of salinity. The long term analysis of groundwater quality demonstrated a deterioration throughout the district.

However, the decline in quality was considerable in north-western areas. The general decline in groundwater quality can be attributed to excessive use of agrochemicals. Exceptionally large deterioration observed in north-western part may be attributed to the lateral intrusion of saline water from southern parts as a result of changes in groundwater movement.

Kaithal district needs judicious management of the groundwater resource. To arrest the declining trend of groundwater levels in critical parts of the district, it is necessary to regulate withdrawals and augment the aquifers with artificial recharge. Moreover, application of modern irrigation techniques like drip irrigation and sprinkler irrigation, are needed to be encouraged.

Techniques such as 'system of rice intensification' (SRI) for growing paddy and the 'alternative wetting and drying' (AWD) can be useful in reducing water use in a rice field by 10-30%. Conjunctive use of groundwater with canal water and proper selection of crops in southern and south-western part can be useful to counter the problem of salinity.

The results also emphasize the need to promote awareness among the end users about water quality to protect aquifers and to implement proper management strategy for sustainability of the resource. Strict regulations are needed to be implemented to reduce the risk of groundwater contamination by indiscriminate use of fertilizers and pesticides.

Reviewers: Dr. Omvir Singh, Department of Geography, Kurukshetra University, Kurukshetra, Haryana, India.

Dr. L. Suri Naidu, International Water Management Institute (IWMI) Patancheru, Andhra Pradesh, India.

REFERENCES

Ballukraya, P. N. (1997). Groundwater over-exploitation: A case study from Moje-Anepura, Kolar district, Karnataka. *Journal of Geological Society of India*, 50: 277-282.

Biswas, S. (2003). Groundwater flow direction and long term trend of water level of Nadia district, West Bengal: A statistical analysis. *Journal of Geological Society of India*, 61: 22-36.

Census of India (2011). *Government of India Publication*. https://www.censusindia.gov.in/

CGWB, (2006). Dynamic ground water resources of India. Central Groundwater Board, Ministry of Water Resources, Govt. of India, New Delhi.

CSO (2013). *Government of India publication*. www.cso.gov.bw/

Chae, G. T., Kim, K., Yun, S. T., Kim, K. H., Kim, S. O., Choi, B. Y., Kim, H. S., and Rhee, C. W. (2004). Hydrogeochemistry of alluvial groundwaters in an agricultural area: an implication for groundwater contamination susceptibility. *Chemosphere*, 55:369-378.

Gleick, P. H. (1996). Water resources. In: *Encyclopedia of Climate and Weather,* ed. S. H. Schneider, Oxford University Press, New York, (2): 817-823.

Goyal, S. K., Sethi, G. K. and Chaudhary, B. S. (2006). Integrated groundwater studies in Kaithal district of Haryana, using remote sensing and GIS. *Proceedings of National Conference on Environmental Conservation.* Sept. 1-3, Pilani, India. 745-751.

HARSAC (2004). Hydro-geomorphological mapping and delineation of groundwater potential zones using satellite data district Kaithal. *Technical report, Haryana Space Application Centre, Hisar: HARSAC/TR/11/04.*

Imtiyaz, M. and Rao, D. J. M. (2008). Influence of over-exploitation on groundwater ecosystem in hard rock terrain. *Proceedings, International Groundwater Conference,* March 19-22, Jaipur, India, 88.

Jain, R. C. (2009). Emerging challenges for sustainable groundwater management in India. *Journal of applied hydrology*, 22(1): 93 101.

Kaur, S., Aggarwal, R., Thaman, S., and Kumar, M. (2009). A micro-level assessment of water resources – a case study. *Journal of Indian water resources society*, 20(1):17-21.

Narain, V. (1998). Towards a new groundwater institution for India. *Water policy*, 1(3):357-363.

Prabhakaran, N., Gurugnanam, B., Vasudevan, S., and Joseph Santhanaraj, S. (2009). Irrigation water quality status studies through GIS in upper Manimukha sub-basin, Villupuram District, Vellar basin, Tamil Nadu, India. *Natural Environment and Pollution Technology*, 8(2):257-260.

Rodell, M., Isabella, V. and James, S. F. (2009). Satellite-based estimates of groundwater depletion in India. *Nature*, 460: 789.

Rotaru, A. and Răileanu, P. (2004). The importance of hydrogeological analyses of groundwater behavior in the slope stability analyses. *Ovidius University Annals Series: Civil Engineering,* 1(6):18-25.

Sarkar, R., Kumar, S., Kumar, Y., and Sharma, H. C. (2007). Groundwater modeling for prediction of water table depth in Ramganga-Bahgulinterbasin of Uttar Pradesh. *Hydrology Journal*, 30(1, 2): 265-278.

Sharma, A. (2008). *Groundwater information booklet Kaithal district, Haryana.* Central Groundwater Board, Chandigarh.

Singh, M. (1992). *Geohydrological conditions and scope of groundwater development in district Kaithal (Haryana).* Groundwater cell, Dept. of Agriculture, Kurukshetra.

Sinha, A. K., Singh, R. P. and Kumar, A. (2006). Groundwater status in Dhund basin, Jaipur district, Rajasthan. *Journal of Geological Society of India*, 68: 809-818.

Statistical abstract (2013). State statistical abstract of Haryana (2007-08). Govt. of Haryana, Chandigarh, India, 910.

Survey of India topographic maps. Government of India publication. www. portal.gsi.gov.in/

U.N. (2006). *The millennium development goals report.* United Nations Department of Economic and Social Affairs, New York: 1-32.

Yadav, M., Sharma, M. P., Prawasi, R., Pal, O., Kumar, R., and Hooda, R. S. (2008). Monitoring area under saathi (summer) paddy in Haryana using satellite data (2003 and 2008). *Technical report, Haryana Space Application Centre, Hisar: HARSAC/TR/03/2008.*

In: Groundwater
Editor: Segun Michael Adelana
ISBN: 978-1-63321-759-1

Chapter 5

GROUNDWATER IN SERBIA: RESOURCES, ENVIRONMENTAL IMPACTS AND SUSTAINABLE MANAGEMENT

Ranko Dragović[1,*], Milan Đorđević[1], Nataša Martić-Bursać[1], Snežana Dragović[2], Ljiljana Janković-Mandić[2], Aleksandar Radivojević[1] and Ivan Filipović[1]

[1]University of Niš, Faculty of Science and Mathematics, Department of Geography, Niš, Serbia
[2]University of Belgrade, Vinča Institute of Nuclear Sciences, Belgrade, Serbia

ABSTRACT

In Serbia, complex geological and geotectonical conditions influenced heterogeneity of hydrogeological structures. The territory of Serbia is divided into five regional geotectonic units which significantly vary depending on the types of water-bearing media. The lower part of Pannonian Basin includes abundant aquifers of Lithothamnian and Sarmatian limestones containing highly mineralised thermal waters, while its upper part includes a number of artesian horizons. In Dinarides,

[*] Corresponding author email: dragovicr@pmf.ni.ac.rs.

groundwater flows through the system of karst channels and fractures and is discharged by high-flow springs. The Carpatho-Balkanides in the eastern part of Serbia are characterised by the masses of Jurassic and Cretaceous limestones, which are heavily fractured and karstified, forming a very permeable medium of diffuse flows drained by numerous springs. The Vardar Zone, which consists of a composite assemblage of continental and oceanic-derived units, is characterised by the poverty in groundwater. In the Serbian Macedonian Mass water-bearing media are represented by Mio-Pliocene sands and coarse grained alluvial deposits of groundwater, but abounding in the occurrence of thermomineral waters. It is estimated that groundwater in Serbia provides about 70% of the demand for water for domestic and industrial purposes. The currently used resources are about 30% of the groundwater potential. This chapter reviews the types of groundwater resources in Serbia, their exploitation, the monitoring of their chemical characteristics, environmental impacts and current activities at national level towards sustainable management.

Keywords: Environmental factors, groundwater, aquifers, chemical characteristics, sustainable management, Serbia

INTRODUCTION

Groundwater is an essential natural resource since most of the Earth's freshwater are stored underground in aquifers. These aquifers provide the valuable baseflow supplying water to rivers during periods of no rainfall. Globally, groundwater is estimated to provide about 50% of current potable water supplies (Table 1), 40% of the demand of self-supplied industry, and 20% of water use in irrigation (UNESCO, 2003). Using current trends and explorative models, Rosegrant and Cai (2002) predict global groundwater withdrawals to increase from 817 km^3/year in 1995 to 922 km^3/year by 2025 (Rosegrant and Cai, 2002).

Table 1. Estimated use of drinking water supply obtained from groundwater (Sampat, 2000)

Region	Use (%)	Population served (millions)
Asia-Pacific	32	1000-2000
Europe	75	200-500
Central and South America	29	150
USA	51	135
Australia	15	3

There are many reasons why society has found it so useful to develop groundwater, but among the most important are (UNEP, 2003):

- aquifers are very convenient sources of water because they are natural underground reservoirs and can have an enormous storage capacity, much greater than even the largest man-made reservoirs;
- many aquifers are also able to offer natural protection from contamination so untreated groundwater is usually cleaner and safer than its untreated surface water equivalent;
- groundwater is relatively easy and cheap to use. It can be brought on-stream progressively with little capital outlay and boreholes can often be drilled close to where the water supply is needed;
- it is a resource that is organizationally easy to develop; individuals can construct, operate and control their own supply, often on their own land.

Geological structure of the territory of the Republic of Serbia has caused the formation and existence of a number of groundwater accumulations which quality and quantity varies in a wide range. Regime and the balance of the water depend on the climate regime and flow regime of rivers. In this chapter the types of groundwater aquifers among different hydrogeological regions of Serbia, the groundwater exploitation and environmental factors affecting its quality are reviewed. The current activities at national level towards sustainable management of groundwater resources are also presented.

Geological Overview

In Serbia, complex geological and geotectonical conditions influenced heterogeneity of hydrogeological structures. The territory of Serbia is divided into five regional geotectonic units (Figure 1): Pannonian Basin, Interior Dinarides, Carpatho-Balkanides, Vardar Zone and Serbian–Macedonian Mass, which may significantly vary depending on the types of water-bearing media (Dimitrijević, 1997, 2002). The Pannonian Basin, the northern unit which occupies over 20% of the territory of Serbia, is composed of Tertiary and Quaternary sediments. Its lower part, composed of Miocene and Lower Pliocene rocks, includes abundant aquifers of Lithothamnian and Sarmatian limestones containing highly mineralised thermal waters. The upper part of the Pannonian Basin, composed of permeable Pliocene and Pleistocene sands and gravels, includes a number of artesian horizons. The Dinarides, extended in the western parts of Serbia, consist dominantly of fractured and karstified Mesozoic limestones. This region, which occupies the largest part of Serbia, is greatly different from the neighbouring regions in respect to hydrogeology and geomorphology. Groundwater flows through the system of karst channels and fractures and is discharged by high-flow springs. The Carpatho-Balkanides in the eastern part of Serbia are characterised by the masses of Jurassic and Cretaceous limestones, which are heavily fractured and karstified, forming a very permeable medium of diffuse flows drained by numerous springs. There are numerous uranium mineralisations in the East Serbian sector of the Carpatho-Balkanides. The Vardar Zone is composed of several blocks of diverse composition, geological history and provenience, and it includes characteristic oceanic elements. This zone, located to the west of the Serbian–Macedonian Mass, consists of a composite assemblage of continental and oceanic-derived units, including also Triassic and Jurassic ophiolites. The Vardar Zone is characterised by the poverty in groundwater. Although varied, rocks of this extremely complex geotectonic zone are mainly poorly permeable to virtually impermeable. The Serbian–Macedonian Mass is located between the Vardar Zone in the west, and the Carpatho-Balkanides in the east. Permeable horizons in the northern half of the zone are underlined by crystalline schists, and water-bearing media are represented by Mio-Pliocene sands and coarse grained alluvial deposits of groundwater, but abounding in the occurrence of thermo-mineral waters.

Hydrogeological Units

According to the Water Management Master Plan of Serbia (Institute for the Development of Water Resources Jaroslav Černi, 2001), the hydrogeological units in Serbia have been classified as follows (Figure 2): the Bačka and Banat region (comprising N and E parts of the Pannonian Basin), the Srem, Mačva and Sava/Tamnava Region (SW Pannonian Basin and NW Dinarides), Southwest Serbia (SW Dinarides), West Serbia (W Dinarides), Central Serbia (Serbian-Macedonian Mass and the Šumadija-Kopaonik-Kosovo Zone), and East Serbia (Dachian Basin and Carpathian-Balkan Arch).

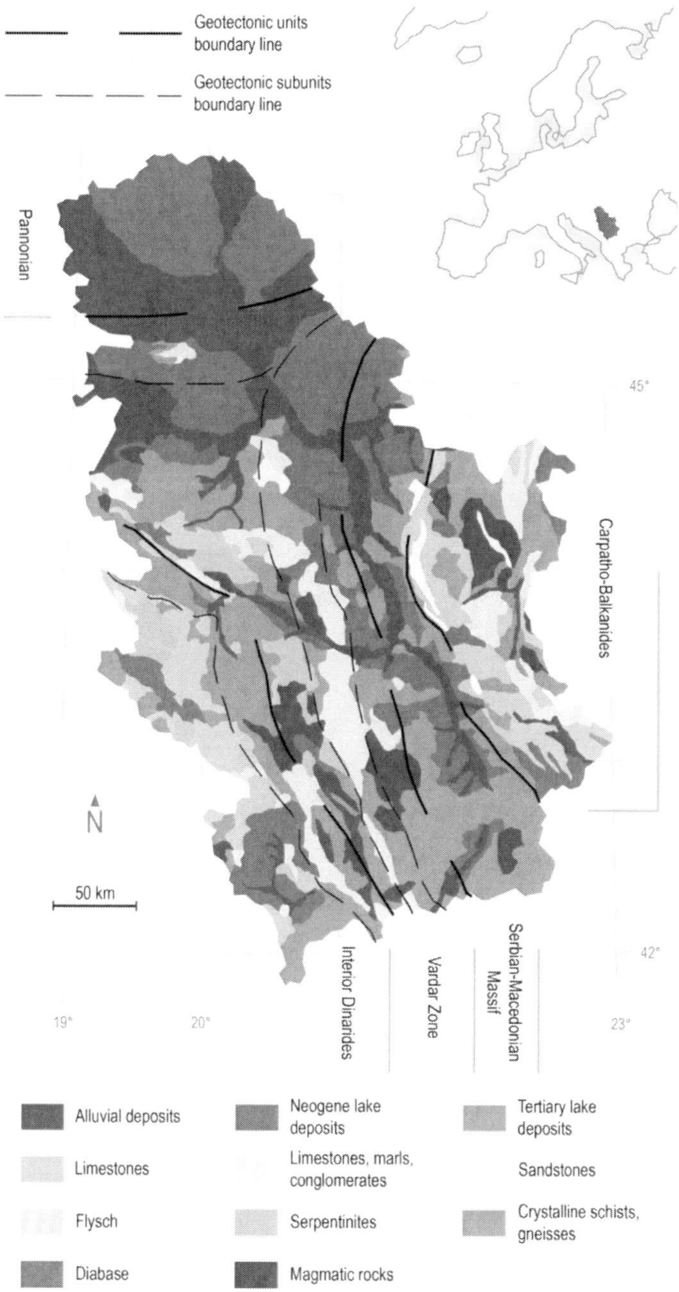

Figure 1. Simplified lithological map of Serbia based on geotectonic divisions (after Dimitrijević, 2002; partly modified).

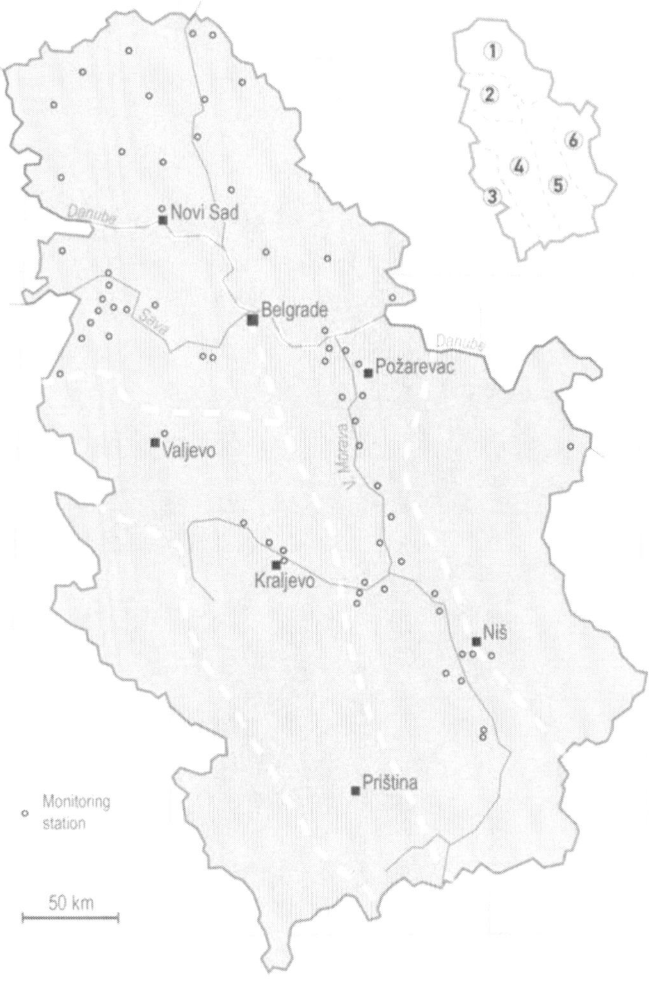

Figure 2. Hydrogeological units of Serbia (1- Bačka and Banat Region, 2-Srem, Mačva and Sava/Tamnava Region, 3- Southwest Serbia, 4- West Serbia, 5- Central Serbia, 6- East Serbia).

The Banat and Bačka Region represents the ultimate southern and south-eastern part of the Pannonian Neogene basin limited to the western and southern side of the Danube. The terrain is composed of Neogene and Quaternary deposits of great thickness (up to 4 500 m) with a number of aquifer environments of different structural types of porosity (fracture and fracture-karst in lower and intergranular in the upper part) and different realms

(Institute for the Development of Water Resources Jaroslav Černi, 2001). The oldest rock masses (shale, granites, serpentines, etc.) represent the lowest floor of the impermeable Pannonian Basin. Directly above chronostratigraphic accumulations are accumulations of mineral water, covered with impermeable rock masses. In the upper parts of the terrain of Bačka and Banat Region, with the exception of central parts of Fruška Gora and Vršac mountains and southeastern Banat, there are Pliocene and Quaternary water-bearing environments. Pliocene water-bearing environments were presented by individual strata of fine- and medium-grained sands, less coarse-grained sands, separated by deeper semi-permeable layers and impermeable rock masses. In these environments, at depths greater than 200 m, aquifers of thermal waters are formed, regularly with greater or lesser quantities of the gases. Quaternary aquifers are present practically from the surface of the terrain to a depth of approximately 230 m. They are presented by polycyclic river and river-lacustrine sandy deposits. These deposits represent basic water-bearing complex in Bačka and Banat which extends to the peripheral parts of Pannonian Basin in Slavonija, Baranja, Hungary and Romania. In water-bearing areas of the basic water-bearing complex sub-artesian and artesian accumulations of groundwater were formed, which represent a major resource for drinking water in the Bačka and Banat Region. In the upper parts of the terrain of Bačka and Banat in sandy gravels, sands, loess and eolian sands to a depth of about 60 m the number of free and artesian aquifer were formed. This groundwater actively participates in the hydrological cycle. Due to high pollution of these aquifers they are only used locally for water-supply. The alluvial aquifer of Danube provides water supply for cities of Novi Sad, Pančevo and Apatin near the Danube. The rate of water abstraction from alluvial aquifers is roughly 2 m^3/s.

The Srem, Mačva and Sava/Tamnava Region is characterized by thick water-bearing alluvial and terrace sediments (up to 30 m) (Dimkić et al., 2011). In the northern part of the region there are no bearing layers, i.e. the accumulations of interest for the exploitation of groundwater for public water-supply of greater settlements (Institute for the Development of Water Resources Jaroslav Černi, 2001). In the southern part of Srem there are important reservoirs of underground water formed in river and river-lacustrine sandy and gravel Quaternary and Pliocene sediments, which are exploited for the public water-supply. In the water-bearing areas of younger Quaternary free subartesian aquifer was formed which is in hydraulic connection with the Sava River. In polycyclic sands and gravels of older Quaternary (the basic water-bearing complex) of 15-50 m thick 2-3 subartesian-artesian aquifers are formed, separated by semi-permeable layers of alevrites and alevrite clay. In Pliocene sands and gravels that represent individual water-bearing environments mutually connected sub-artesian and artesian aquifers are formed. In the area of Mačva, primary water-bearing environment is presented by gravel and sandy-gravel sediments of younger and older Quaternary which are interconnected over the area. Sub-artesian and artesian aquifers in the area of Mačva were formed in the complex of sandy and sandy-gravel Pliocene sediments alternating with alevrites and alevrite clays to a depth of about 270 m. In the area of Sava/Tamnava, the water-bearing environments of Quaternary are of uneven distribution and thickness. The Sava alluvium holds a major water source for the City of Belgrade, Serbian capital, with current rate of water abstraction of 4-5 m^3/s (Institute for the Development of Water Resources Jaroslav Černi, 2006).

The region of Central Serbia is noticeably different from the surrounding two units West Serbia and East Serbia in terms of hydrogeology. In this area the water-bearing media are represented by sands of Mio-Pliocene, gravel-sandy sediments of Quaternary and alluviums of the Great, South and Western Morava (Institute for the Development of Water Resources Jaroslav Černi, 2001). Quaternary sandy-gravel deposits in the alluvial plain of the Morava represent the primary water-bearing environment in this area. Groundwater accumulated in the bearing areas are in direct hydraulic connection with river watercourses. There are only few karst aquifer environments in the area. Fissure environments are highly abundant, especially in southern part of this hydrogeologic region. The most important aquifers of the Central Serbia Region are alluvial aquifer of Velika Morava River (6-60 m thick) and that of Danube River (15-30 m thick) (Dimkić et al., 2011).

The East Serbia Region is characterised by highly karstified Upper Jurassic and Lower Crataceous limestones of the Carpathian-Balkan Arch and a portion of the Dachian Basin, with a large number of karstic springs, 16 of which of a minimum yield of 100 L/s (Stevanović and Filipović, 1994). Infiltration of precipitation to permeable rock masses is high. Large alluvial aquifers are rare in this region (the Nišava and Beli Timok basins). From the point of the use of groundwater, unfavorable moment represents high variations of amount of water of most karst springs (Institute for the Development of Water Resources Jaroslav Černi, 2001).

In terms of hydrogeology the most important formation in this area are limestones of the Middle and Upper Triassic (Institute for the Development of Water Resources Jaroslav Černi, 2001). They are characterized by a large thickness in the southern half of the region, where the area from Pešter plateau to northern Metohija basin represent a more or less uniform limestone mass. Karstification in the tectonically damaged limestone environment occurred to a considerable depth, and the main conductors of groundwater represent diverse karst forms. The area of karst is characterized by very high infiltration of atmospheric precipitation. The movement of groundwater takes place through a system of karstic channels and expanded fissures up to the lowest drainage bases, where the strong karst springs are formed. Among numerous karstic springs, eleven have a minimum discharge of more than 1 m^3/s of water (Stevanović and Filipović, 1994; Stevanović, 1995).

In the region of West Serbia the most important hydrogeological formation is composed of extensively karstified Middle and Upper Triassic limestones (Dimkić et al., 2011). In comparison to other hydrogeological regions this area is poor in groundwater (Institute for the Development of Water Resources Jaroslav Černi, 2001). Significant amounts of the groundwater are accumulated only in sandy-gravel layers of main water courses. The alluvial aquifer is represented by the rivers the Drina, Kolubara, Western Morava and Ibar. Lacustrine sediments of Neogene age were deposited in several relatively spacious basins, but are generally characterized by low permeability. Low permeable fissure area covers the majority of West Serbia Region. There are sporadic aquifers formed in serpentinites and perioditites while the other environments are marked by numerous sources of yield below 0.001 m^3/s.

GROUNDWATER EXPLOITATION

About 75% of water for public water supply in Serbia is abstracted from groundwater resources. Based on Water Management Master Plan of the Republic of Serbia (Institute for the Development of Water Resources Jaroslav Černi, 2001) about 23 m^3/s of groundwater is currently used (Table 2).

Based on up-to-date research, the current yield of groundwater sources is roughly 30% out of the groundwater potential of 67 m^3/s (Table 3) which could be used in the future (Dimkić et al., 2011). The largest groundwater reserves exist in alluviums near the confluences of powerful watercourses and within karstic aquifers. According to the Water Management Master Plan of Serbia and based on current information, artificial recharge can provide an additional 40 m^3/s of water (Institute for the Development of Water Resources Jaroslav Černi, 2001).

Table 2. Groundwater source yield in Serbia by type of water-bearing medium (m^3/s) (Institute for the Development of Water Resources Jaroslav Černi, 2001)

Region	Alluvial aquifers	Quaternary subartesian aquifers	Neogene aquifers	Karstic aquifers	Fissured aquifers	Total
Bačka and Banat	1.45	3.57	0.43	-	-	5.45
Srem, Mačva, Sava/Tamnava	6.97	0.34	0.51	0.03	-	7.85
Central Serbia	2.58	-	0.84	0.43	-	3.85
East Serbia	0.62	-	0.06	1.71	-	2.39
Southwest Serbia	0.24	-	0.14	1.61	-	1.99
West Serbia	1.05	-	0.06	0.40	0.02	1.53
Total	12.9	3.91	2.04	4.18	0.02	23.1

Table 3. Estimated groundwater potential in Serbia (without artifcial recharge) (m^3/s) (Institute for the Development of Water Resources Jaroslav Černi, 2001)

Region	Alluvial aquifers	Quaternary subartesian aquifers	Neogene aquifers	Karstic aquifers	Fissured aquifers	Total
Bačka and Banat	9.39	4.91	0.55	-	-	14.8
Srem, Mačva, Sava/Tamnava	21.1	0.55	0.99	0.10	-	22.7
Central Serbia	9.93	-	1.73	1.48	0.18	13.3
East Serbia	1.06	-	0.24	2.98	-	4.28
Southwest Serbia	0.57	-	0.33	7.28	-	8.18
West Serbia	1.73	-	0.12	1.89	0.03	3.77
Total	43.8	5.46	3.96	13.7	0.21	67.1

Until now, about 70 karstic sources with the total minimal capacity exceeding 4.5 m^3/s have been tapped for centralized water supply (Stevanović et al., 2007). Most of the large cities of Eastern and Western Serbia currently use karst groundwater for water supply (Niš, Bor, Pirot, Paraćin, Ćuprija,

Valjevo, Novi Pazar, Prijepolje, Peć, Priboj, etc.). This number does not include the tapped springs for villages, smaller settlements, and factories.

However, due to unstable flow regime when only natural spring flow is tapped, numerous problems arise during the recession period (summer–autumn). In addition to natural groundwater potential, additional amounts of groundwater can be obtained by artificial recharge methods, from water sources in alluvial plains and by controlling karstic springs. During the last two decades, several newly constructed systems for artificial control of karst aquifers (mostly in East Serbia) resulted in significantly improved water supply.

Water supply of rural settlements is done almost exclusively by the groundwater water. Approximately 6.2 m^3/s of groundwater is used for water supply for the population in rural settlements in Serbia (Institute for the Development of Water Resources Jaroslav Černi, 2001).

Complex geological structure and favorable hydrogeological relations created on the territory of Serbia significant wealth in mineral and thermal waters, with about 1200 registered occurrences (Institute for the Development of Water Resources Jaroslav Černi, 2001). The total yield of springs with temperatures above 20° C, for the part of the country without provinces, is about 2.0 m^3/s, which corresponds to is the energy of $3 \cdot 10^{15}$ J/year. It is estimated that only small part of the registered thermal water potential is used (about 1% of renewable reserves). Until today the use is practically reduced to health and balneological needs and bottling (potable water).

Use of water for balneological and recreational purposes is traditionally the most important aspect of use of mineral and thermal waters. In Serbia, this type of groundwater is used in 49 active natural spas and resorts, with developed balneological, recreational and tourism industry (Institute for the Development of Water Resources Jaroslav Černi, 2001).

Among the many registered occurrences of mineral water only small part is used in industrial production for bottling. Today in Serbia, total annual production is of about 150 million liters of water and about 50 million liters of soft drinks (Institute for the Development of Water Resources Jaroslav Černi, 2001).

The terrains of Serbia provide great opportunities for the use of hydro energy both because of the amount of hot water that can be obtained and the relatively small depth at which collectors are often situated. The use of geothermal energy for heating and other energy purposes is practically at an early stage and it is very modest compared to the potential of these resources. In the hydrogeothermal system of the Pannonian basin 23 boreholes are capped with a total heat output of around 25 MW (Institute for the Development of Water Resources Jaroslav Černi, 2001). The total installed thermal power at localities south of the Sava and Danube rivers is 74 MW.

ENVIRONMENTAL FACTORS AFFECTING THE GROUNDWATER QUALITY

Groundwater degradation occurs in the cases of excessive exploitation, inappropriate or uncontrolled activities at the land surface, including disposal of waste and spillage of chemicals, which contaminate the underlying aquifer and major change of land use (UNEP, 2003). The nature of the aquifer will also influence the scale of the contamination problem. Thus, in a highly fractured aquifer where groundwater flow is easy and relatively rapid, contamination may become more widely dispersed in a given time than where flow is intergranular.

There is considerable diversity of groundwater quality in Serbia, ranging from waters of high quality and across the waters that before use, and depending on the application, must be subjected to standard treatment to the waters that require a complex processing procedure. Groundwater from basic water-bearing complex in Vojvodina, the Serbian north province, contains high baseline concentrations of organic matter, arsenic, iron and manganese (Institute for the Development of Water Resources Jaroslav Černi, 1997/1999). In some areas, inadequate protection resulted in water quality changes due to anthropogenic impacts. In the alluvium of the lower part of Velika Morava River valley high nitrate concentrations were observed as a consequence of uncontrolled wastewater discharges and agricultural practices (Boreli-Zdravković and Damjanović, 2007). The study comprising 66 different key sampling sites in ten representative sites in Serbia indicated the presence of severe pollution of groundwater, with 35% of the studied wells being contaminated by Mn, As and NO_3 (Devic et al., 2014). The arsenic concentrations in most samples in Banat (25.7– 137.8 μg/L) and sample at the Great Morava River (21.7 μg/L) exceeded the World Health Organisation (WHO) recommended value of 10 μg/L for drinking water (WHO, 2008). The region with the highest trace metal pollution by lead (15.8 μg/L), nickel (41.11 μg/ L), and cadmium (0.8 μg/L) is located in Krusevac City, which is exposed to large uncontrolled metal inputs from anthropogenic sources (Devic et al., 2014). The groundwater pollution by heavy metals (iron, zinc and lead) has also been observed in the Ibar alluvium near Raška as a result of an ore flotation facility (Miladinović et al., 2012).

Significant quantities of waste water and other material from settlements, industry and agriculture, which are discharged into water streams and groundwater without prior treatment, cause degradation of water resources in Serbia as a whole. According to the degree of natural protection slightly better situation is with the basic water-bearing complex in Vojvodina and sandy Neogene aquifers (Institute for the Development of Water Resources Jaroslav Černi, 2001).

Karst aquifer environments are open to direct pollution from the surface and aquifer environments in the alluvium by pollution both from the surface and from polluted river flows. In addition to pollution processes, uneconomical use of groundwater produced negative effects in many areas (primarily Vojvodina). South of the Sava and Danube rivers pollution process affected a significant part of groundwater reserves. In addition, some of the water potential is disturbed by exploitation of gravel, followed by raising industrial facilities and urbanization. Among groundwater, those in the alluvium areas are the most exposed to the pollution, particularly through the impact of polluted surface waters and from the terrain surface (fertilizers, petroleum products, cesspits, etc.).

Groundwater is somewhat enriched in respect to radium when compared with surface waters especially in mineralised areas. The occurrence and distribution of radium in groundwater is strongly influenced by geological and hydrogeological conditions of the area, although the different rock complexes of the same geotectonic unit could display wide ranges of radium concentrations. From the Late Crataceous until the Miocene the intensive volcanic activity took place on the territory of Serbia, where large masses of diverse effusive products were deposited, although uranium mineralisations occurred in older deposits (Jelenković, 1999). As uranium mineralisations have been created under different physical and chemical conditions they are present in different modes of occurrence and in a wide range of geological areas. In granitoid and volcanic complexes of Serbia uranium occurs in its own minerals formed during magma crystallisation, in the accessory minerals of granitoid complexes, and in minerals formed as a consequence of hydrothermal activities. The volcanic complexes which spread over a significant part of the territory of Serbia are characterised by anomalous uranium concentrations. The majority of occurrences and deposits of uranium are associated with clastic sedimentary rocks, e.g. fine-grained sandstones with high organic matter content, particularly in basins close to granitoid complexes. Different uranium mineralisations also occur in metamorphic complexes. The remobilisation of uranium occurred in most types of rocks in the territory of Serbia, indicating the post magmatic redistribution of uranium had taken place. A survey of radium activity concentrations in well and spring waters from 170 sampling sites in Serbia (Dragović et al., 2012) indicated that 90% of ^{226}Ra activity concentrations were below 1 Bq/L, which is the guidance level recommended by WHO (WHO, 2008). The highest activity concentrations of this radionuclide (up to 17 Bq L^{-1}) were found in crystalline rocks- and carbon dioxide rich-aquifers. The mean activity concentrations of ^{226}Ra were found to be 0.36 Bq/L and 0.57 Bq/L in well and spring waters, respectively. Their spatial distribution is shown in Figure 3.

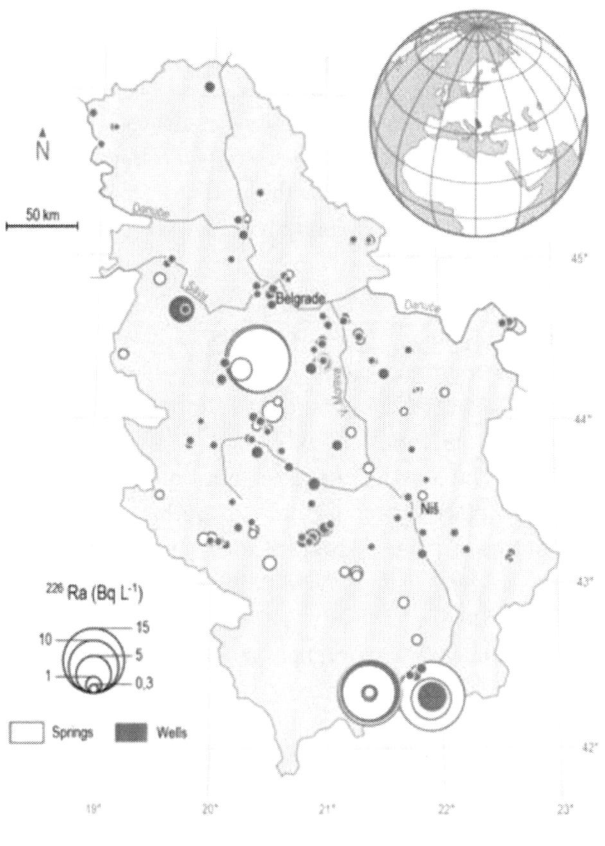

Figure 3. Spatial distribution of the ^{226}Ra activity concentrations in the well and spring waters in Serbia. Reprinted from Dragović et al., 2012, with permission from Elsevier.

Sustainable Management of Groundwater

The effective management of groundwater resources requires awareness of the status of groundwater – both its quality and quantity available, understanding of the aquifer sufficient to be able to identify options to remedy a problem situation, water laws, surveillance and awareness in governmental planning and society of the importance of groundwater (UNEP, 2003). The monitoring of water quality and water levels provides information necessary for rational water management on all kinds of water sustainability issues, i.e. understanding the flow system and the baseline water quality, identifying problems of local overdraft or water pollution, providing independent information on the rate of use of the resource and evaluating the effectiveness of management actions.

Groundwater resources of Serbia are monitored at national, municipal and water supply source level, as well as in a portion of the riparian lands of Danube, Sava and Tisa rivers which are within the reservoir zone of the Iron Gate Dam. The network of monitoring stations is established under the responsibility of the Hydrometeorological Service of Serbia to monitor the quantity and quality of surface water and groundwater. Groundwater quality is being systematically monitored through an alluvial aquifer monitoring network, but systematic monitoring of Neogene and karstic aquifers has not been yet established (Figure 2). About 50 parameters are analyzed once per year: the main cations and anions, nitrogen components, heavy metals, pesticides etc. Due to inadequate distribution of sampling sites information on groundwater quality in large parts of Serbia are very scarce or missing. The extent and substance of current monitoring activities in Serbia are not appropriate for comprehensive groundwater quality and quantity monitoring. It is, therefore, necessary to upgrade current monitoring network to encompass both shallow aquifers and deep karst and Neogene aquifers, through the inclusion of groundwater user facilities and the establishment of new monitoring sites.

The need for an integrated approach to monitoring design and resource management becomes even more important when dealing with transboundary aquifers, where the complexity of geology, climate and human activities that defines a groundwater catchment if further complicated by political and legal differences of two or more neighbouring countries. In recent years, increasing attention has been paid to the identification, assessment and management of transboundary aquifers. In Serbia, some of these activities were synchronized through direct bilateral communication of national institutions (Đurić, 2006), others were undertaken by United Nations-coordinated action (UNECE, 2007), while some resulted from European Union funded projects (Stevanović et al., 2011). The implementation of joint transboundary aquifer management principles includes sustainable exploitation of groundwater of neighbouring countries.

Major tasks in the area of groundwater in Serbia are related to groundwater legislation, preparation of studies and planning documents, development of groundwater monitoring and information system, development of regional groundwater resources and achievement of conditions for sustainable use of groundwater resources (Dimkić et al., 2011). The priorities in groundwater protection remain: protection of surface water which recharges groundwater, regulated wastewater treatment, agricultural production harmonised with groundwater protection requirements and regulation of water source protection zones.

Conclusion

Serbia is relatively rich in groundwater, deposited mainly in thick Quaternary and Neogene intergranular aquifers. Despite of their relatively high abundance and intensive use, groundwater in Serbia, is not sufficiently investigated. The information on their resources, the extent of exploitation, quality and regime is still missing in some areas. The monitoring network should be upgraded to comprehensively address both quantitative and qualitative status of groundwater aquifers, since only accurate data would enable proper water resources assessment and their sustainable use. The principles of sustainable use and protection across Serbia should be adopted as a pre-requisite for the improvement of groundwater monitoring system.

Acknowledgments

This work was supported by the Ministry of Education, Science and Technological Development of the Republic of Serbia (Project No. III43009). The authors are grateful to Prof. Jovan Romelić, Faculty of Sciences, University of Novi Sad for his careful reading of the text and helpful remarks.

REFERENCES

Boreli-Zdravković, Đ., Damjanović, M. (2007). Current status and problems associated with alluvial waters in the Velika Morava River valley, Regional IWA Conference on Groundwater Management in the Danube River Basin and Other Large River Basins, Belgrade, Serbia, 7–9 June 2007, Proceedings, 201-208.

Devic, G., Djordjevic, D., Sakan, S. (2014). Natural and anthropogenic factors affecting the groundwater quality in Serbia, *Science of the Total Environment* 468-469, 933-942.

Dimitrijević, M.D. (Ed.) (2002). Geological Atlas of Serbia, Serbian Ministry of Natural Resources and Environmental Protection, Belgrade, 187 pp.

Dimitrijević, M.D. (1997). Geology of Yugoslavia. Geoinsitute-Barex, Belgrade.

Dimkić, M, Stevanović, Z., Đurić, D., 2011. Progress and improvement of the status of groundwater in Serbia, IWA Specialist Groundwater Conference, 8-10 September 2011, Belgrade, Serbia, Proceedings, 81-101.

Dragović, S., Janković-Mandić, Lj., Dragović, R., Đorđević, M., Đokić, M. (2012). Spatial distribution of the ^{226}Ra activity concentrations in well and spring waters in Serbia and their relation to geological formations, *Journal of Geochemical Exploration* 112, 206-211.

Đurić, D. (2006). Transboundary groundwater bodies - basis for shared source management, Conference "Water 2006", 6-9 June 2006, Zlatibor, Serbia, Proceedings, 7-12.

Institute for the Development of Water Resources Jaroslav Černi (2001). Water Management Master Plan of the Republic of Serbia, Belgrade.

Institute for the Development of Water Resources Jaroslav Černi (2006). Study of Belgrade's Water Source: Present State and Directions of Development, Phase 1, Study, Belgrade.

Institute for the Development of Water Resources Jaroslav Černi (1997/1999). Design Concept for Water Supply in Vojvodina, Phase 1 and Phase 2, Study, Belgrade.

Jelenković, R. (1999). Deposits of Metallic Orres, Faculty of Mining and Geology, University of Belgrade, Belgrade (in Serbian).

Miladinović, B., Papić, P., Mandić, M. (2012). Groundwater pollution with heavy metals in the Ibar alluvium near Raška (Serbia), *Annales Géologiques de la Péninsule Balkanique* 73, 125-130.

Rosegrant, M.W., Cai, X. (2002). Global Demand and Supply Projections, Part 2: Results and Prospects to 2025, *Water International* 27, 170-182.

Sampat, P. (2000). Deep trouble: the hidden threat of groundwater pollution. *Worldwatch Paper*, No. 154, Washington DC: Worldwatch Institute.

Stevanović, Z. (1995). Karst groundwater in Serbia, In: Water Mineral Resources of litosphere of Serbia (ed. Stevanović, Z.), Special Edition, Faculty of Mining and Geology, Belgrade, pp 77-120 (in Serbian).

Stevanović, Z., Filipović, B. (1994). Hydrogeology of Carbonate Rocks of the Carpatho-Balkanides. In: Groundwaters in Carbonate Rocks of the Carpathian-Balkan Mt Range, (eds. Stevanović, Z., Filipović, B.), Special Edition of CBGA, Allston, Jersey, pp. 203-237.

Stevanović, Z., Jemcov, I., Milanović, S. (2007). Management of karst aquifers in Serbia for water supply, *Environmental Geology* 51, 743-748.

Stevanović, Z., Kozák, P., Lazić, M., Szanyi, I., Polomčić, D., Kovács, B., Török, J., Milanović, S., Hajdin, B., Papić, P. (2011). Towards sustainable management of transboundary Hungarian-Serbian aquifer. In: Transboundary Water Resources Management. A Multidisciplinary Approach, 1st edition (eds. Ganoulis, J., Aureli, A., Fried, J.), Wiley-VCH, Verlag, pp. 143-149.

UNECE (*United Nations* Economic Commission for Europe) (2007). Our waters: joining hands across borders – first assessment of transboundary rivers, lakes and groundwaters, UNECE, Geneva.

UNEP (*United Nations Environment Programme*) (2003). Groundwater and its susceptibility to degradation: A global assessment of the problem and options for management, UNEP/DEWA/RS.03-3.

UNESCO (*United Nations Educational, Scientific and Cultural Organization*) (2003). The UN World Water Development Report. Retreived from http:// www.unesco.org/water/wwap/wwdr/wwdr1/pdf/chap4.pdf

WHO (World Health Organization) (2008). Guidelines for Drinking-water Quality, Vol. 1, Recommendations, Geneva, 197-209.

Reviewed by Prof. Jovan Romelić, Faculty of Sciences, University of Novi Sad.

In: Groundwater
Editor: Segun Michael Adelana
ISBN: 978-1-63321-759-1

Chapter 6

SUSTAINABLE GROUNDWATER RESOURCES MANAGEMENT FOR LANGAT BASIN, SELANGOR, MALAYSIA THROUGH NUMERICAL MODELLING

Saim Suratman*
Hydrogeology Research Centre, National Hydraulic Research Institute of Malaysia, Selangor, Malaysia

ABSTRACT

Rapid and ultra-large scale development is progressing in the national capital of Kuala Lumpur and the adjoining areas in Selangor State, Malaysia and has somewhat put greater pressure on water demands. It is understood that the southern parts of Selangor will experience a degree of water shortages due to the unforeseen circumstances in relation to the transfer of water from Pahang to Selangor. To cope with the water deficit in Selangor State, groundwater is considered as the most viable supplementary water source to fill the gap in demand for the public water supply system during the period in which the surface water is unable to fulfill the requirement. Most of the groundwater in Selangor is used for industrial purposes and currently, some factories are already started

* Corresponding author: Saim Suratman. Hydrogeology Research Centre, National Hydraulic Research Institute of Malaysia, Lot 5377, Jalan Putra Permai, 43300 Seri Kembangan, Selangor, Malaysia. E-mail: saim@nahrim.gov.my.

shifting their water source from surface water to groundwater. In Selangor, the yearly abstraction of groundwater in 2005 from licensed wells was estimated at 1.2 million m^3 and increased to 13 million m^3 in 2008 and 20 million m^3 in 2010.

Groundwater simulation with the overall objectives of estimating groundwater resources potential in the Langat Basin area and to predict environmental impacts caused by groundwater abstraction was previously carried out by JICA in 2002. In the modelling work, a number of application scenarios have been carried out. The results of nine selected scenarios (Simulation Variants) characterizes the groundwater conditions in the study area and potentially enable to support technical decisions and policy making. The modelling was carried out again recently, with additional newly acquired data to improve the previous findings.

Keywords: Groundwater, modelling, MODFLOW, sustainable, Langat Basin, Malaysia

INTRODUCTION

The previous groundwater flow modelling was utilizing the integrated mathematical modelling package of the three-dimensional finite-difference Visual MODFLOW Version 2.8.2 (McDonald and Harbaugh, 1988; Guiguer and Franz, 1997) whereas the recent modelling exercise was using Version 2010.1.

For the purpose of managing the resource sustainably, the groundwater modelling was carried out in the area to:

i improve the understanding of the modelled area (proposed abstraction area) and its surrounding regional groundwater flow system from the idealization of the aquifer system in this area;

ii predict the steady-state flow characteristics and the future regional changes caused by variable discharge of groundwater in the proposed abstraction area following extensive abstraction of groundwater from the multi-layered aquifers;

iii provide a calibrated model able to be utilized in the future study; and

iv recommend the sustainable volume of groundwater that could be abstracted from the Langat Basin.

The regional groundwater modelling work is concentrated in the alluvial part of the Langat Valley and the domain was selected according to natural hydrogeological boundaries, over an area with *x* coordinates 378000 – 418000 m, and *y* coordinates 285000 – 330000 m (area 40 x 45 km).

Study by JICA (2002), Suratman (2005) recommended that 45,000 m^3/day of groundwater could be sustainably abstracted from the alluvial aquifers in the Langat Basin. An earlier modelling was done by Suratman (1999).

However, considering taking 5% of the recharge, the available groundwater to be tapped sustainably is at 130,000 m^3/day.

Utilizing the groundwater model, it assists is able to help in ensuring the sustainability of the groundwater utilization from the alluvial aquifers in the Langat Basin.

The National Economic Action Council (1998) has identified groundwater as one of the highly potential water resources to be developed in the area.

Topography, Geology and Hydrogeology of the Area

Introduction

The quality of modelling results depends very much on the availability and quality of various data.

This data includes instructions data for setting up the model, which includes river cross-sections, surface topography, geological layer boundaries, as well as information data for calibration and validation of the model, which are time series of measured groundwater levels in monitoring wells, water levels in rivers, and abstraction of groundwater from wells, rainfall, and evapotranspiration. The painstaking compilation of existing data within the project area shows that it is relatively good in terms of quantity and quality, for model set up, calibration and verification. Hence, the current field investigation, that includes development of new wells in Tg. Duabelas and Labohan Dagang, as well as the measurements and collection of various additional data was performed to support the required modelling works.

A number of previous works conducted within the basin include Abdul Rashid and Mohammed Hatta (1998, 2000), Drillco Technologies (M) Sdn. Bhd. (2005), Mohamed Hatta and Abdul Rashid (1999), JMG (2005) and Aros (2009).

Surface Topography

The topographical information is the most important for the construction of the physical model boundary. The surface topography of the modelling area was prepared using the available digital topographic maps.

Using GIS software (ArcGIS), a digital elevation model, especially for in the central part of Langat Basin area was created, and the model was used for constructing the groundwater model.

Topographically, the major surface area in the Study area is covered by peat and peaty soils, which are of high plasticity and have has high water bearing capacity and contain a large portion of organic matter. These peat layers act to store the excess rainwater and are only drained by a shallow surface drainage system. The surface topography and configuration of drains determine the directions of groundwater flow in the shallow groundwater horizon.

Geology and Aquifer Characteristics

Geological Units

There are 5 geological units encountered in the Study Area (Hussin, 2001; JICA, 2002). They are:

1 Beruas Formation, consisting of terrestrial sediments (clayey/silty peat and peat with decayed wood and roots) deposited during the later Holocene period. The presence of organic materials has caused the acidity and brown colour of the peat and surface water. This formation is up to few meters thick.

2 Subsurface Gula Formation consists of grey to greenish grey Holocene clay, silt and sand with shells, detritus remains and a minor amount of gravel deposited under estuarine and shallow shelf marine environments (aged at approximately 10,000 years). The Gula Formation acts as a nearly impermeable layer confining the formations beneath it.

3 Kempadang Formation (Upper clay member of Simpang Formation) – The formation consists of Late Pleistocene age clay, silt and sand deposited in shallow marine environment (approximately 0.6-0.12 million years). The clayey sediments of Kempadang Formation are considered as very low permeable. The Gula and Kempadang

Formations are of similar composition and therefore are generally described as one unit. The distribution area of Kempadang Formation is uncertain and thought to be limited.

4 Lower sand member of Simpang Formation – The main aquifer consists of terrestrial Pleistocene (10,000-1.8 million years) deposits made up of largely sand and gravel with minor clay and silt content. Alternatives of sand/gravel layers and clay layers are common futures. The coarse and very coarse sand and gravel occurred at the base of the formation.

5 Bedrock – Pre-Quaternary weathered hard-rocks like sandstone, quartzite, shale, limestone, schist, slate, mudstone, or granite (mainly Kenny Hill, Kajang, or Hawthornden Formation).

Bedrock

Bedrock is the second main topographical aquifer boundary, which is defined in the model as an impervious or low permeable bottom boundary. The topography of the bedrock in the modelling area was interpreted using available topographic maps of terrain, geological logs from existing and new boreholes, and geological profiles available in existing reports.

The bedrock was introduced in the model set up with uniform thickness of 20 m as the model layer No. 7 (bottommost layer). The weathered and to some degree permeable bedrock enables infiltration of rainwater and further aquifer recharge in the zone of the exposed outcrops.

Topography and Properties of Aquifer Layers

The topography of the first peat layer follows the topography of the surface and was modelled with thickness of 2 m. This layer inhabits input of net precipitation (rainfall minus evaporation) and surface outflow including run-off in drains or canals. The topography of underlying layers (2 and 3) is based on the evaluation of geological logs from existing boreholes.

The reliable data for modelling work were obtained from various comprehensive evaluations of geological logs, soil investigation, and pumping tests results performed previously in the Study area. Aquifer mapping in the area using geophysical methods were carried out by Umar et al. (2000) and Ibrahim et al. (2001, 2003a and 2003b).

In the model, the formation was simplified to the actual hydrogeological conditions, consisting of the following four layers:

1 Layer 1 - peat and peaty soil (Beruas Formation)
2 Layer 2 - clayey soil (Gula Formation and Kempadang Formation, when exist)
3 Layer 3 - the Aquifer, made up of sandy soils/gravelly soil with intercalation of clayey soil (Lower Member of Simpang Formation)
4 Layer 4 - bedrock

However, based on the compiled data, a thorough evaluation was made leading to divide the aquifer from 4 layers into 7 different layers (including sub-layers), with the following hydrogeological characteristics:

- Layer 1: Peat, represents subsurface shallow and a thin aquifer with the ability of taking rainwater and transmitting the surplus water as a shallow groundwater flow; in this layer the simulation of surface runoff into the drains, canals, or rivers takes place;
- Layer 2: Silty and sandy clay, upper low permeable to impermeable aquitard of regional extent;
- Layer 3: Silty or clayey sand, usually with a minor amount of gravel, represents the upper less permeable aquifer; this layer is rather heterogeneous in thickness and extent, at some places missing, at other places rather thick and inter-bedded with clayey horizons (Layer 3a);
- Layer 4: Sandy clay or clayey silt, lower low permeable to impermeable aquitard of regional extent (Layer 3b);
- Layer 5: Silty sand and gravel, sometimes with occurrence of layers of sandy clays, is characteristic with higher anisotropy and represents the upper part of the main highly permeable regionally spread aquifer (Layer 3c);
- Layer 6: Silty sand and gravel, represents the lower part of the main highly permeable regionally spread aquifer (Layer 3d); and
- Layer 7: Bedrock, very low permeable or nearly impermeable, weathered, pre-Quaternary rock (Layer 4).

The model vertical discretization is based on lithological composition and hydrogeological properties of sediments through compilation of existing data as well as data derived from the current study.

In the modelling work, Layers 5 and 6 are the most important as they are considered to be the productive aquifers.

The other layers represent the main aquifer boundaries, which affect the recharge characteristics of the area and thus are also important for aquifer management. Previous works by Noor (1980) and Awadalla et al. (1989) in Kelantan estimated that recharge is 12% of the amount of rainfall.

Hydrogeology and Groundwater Flow

The main aquifer is recognized as confined. Groundwater is recharged from rainfall through:

- i Downward flow through the aquitard,
- ii Infiltration of water around the edges of bedrock exposure where the aquitard is thin or absent,
- iii Flow from relatively more permeable bedrock,
- iv Infiltration from riverbed in stretches where the river bottom is in contact with more permeable sandy horizons, and
- v Infiltration from ponds and wetlands where the upper aquifer and the aquitard have been removed and replaced by more permeable materials.

Natural groundwater discharge is considered primarily at coastal areas atto the sea as well to the river and drainage systems.

Confining clayey layers are low in conductivity and leaky. In the past natural conditions, the piezometric levels in main aquifer near the recharge zones were probably higher than the peat water level. The leakage was in an upward direction. At present, in the areas with lower aquifer piezometric level, the leakage is directed from the surface downwards to the main aquifer.

The natural groundwater flow is affected by abstraction of water from existing wellfields and from the active dewatering at former tin mining areas. In the area of the Imuda sand mining pit the lowering of the water table down to the gravels (or bedrock) has taken place (estimated 40 m below the ground surface). Future additional wellfields will create potential change in the groundwater regime in the area.

The regional groundwater flow is characterized by measurements of piezometric heads performed during the Study as well as data compiled from various previous works. The figures show that a depression cone has developed at the Megasteel area and Imuda sand mining pit. The lowering in piezometric level to about –8 m from the mean sea level was observed.

Hydrology

Various hydrological data is important in simulating the groundwater regime in the Study area.

The main surface water body in the area is made up of Langat River and its tributaries (Semenyih, Chuah, Labu and Kelambau).

Mainly shallow agricultural ditches form the drainage, which discharges into the Langat River or into its tributaries.

The collected hydrological data can be summarized as the annual water balance in the area which as follows:

1 Annual average rainfall in the area is 2238 mm/year.
2 Annual average pan-evaporation is 1606 mm/year.
3 Annual average evaporation, estimated as 80% of average pan-evaporation, is 1284 mm/year.
4 Difference between average rainfall and average evaporation is 954 mm/year (effective rainfall).
5 Annual measured average discharge (surface run-off) is 846 mm/year.
6 Annual average Groundwater Recharge was estimated at 108 mm/year.

Annual average groundwater recharge in the upstream Langat River Basin is estimated at 139×10^6 m^3/year and about 5% of rainfall in the area.

Groundwater Abstraction

At present the major groundwater abstraction is carried out on several places (Figure 2.6.1 and 2.6.2). The main well field is located in Megasteel property (wells PWM-1, PWM-2, PWM-3, PWM-4, PWS-1 and TW-4). The smaller capacity well fields are in Brooklands Estate (Kajibumi WF 1, 4 wells) and in Olak Lempit (Kajibumi WF 2 and TU/B).

The natural groundwater flow is affected by abstraction of water from the Imuda tin mine. Estimated and simulated daily pumping rates from well fields and dewatering pits are listed in Table 2.6.1.

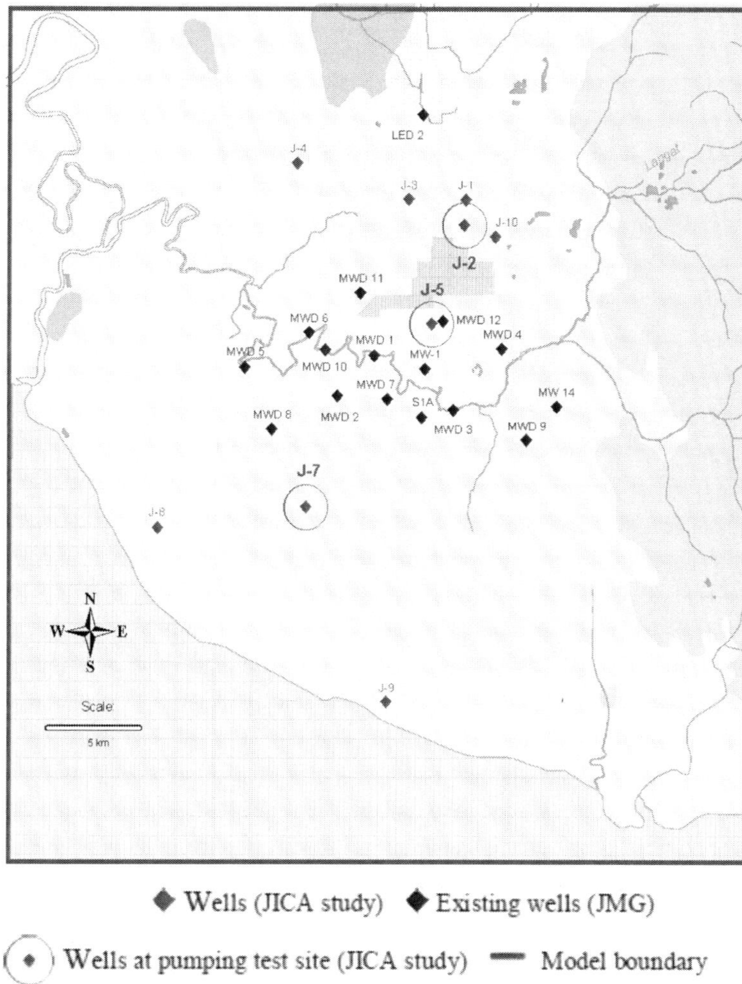

After JICA, 2002.

Figure 1. Monitoring Wells and Pumping Test Sites of the previous investigation.

Groundwater Flow Model

Modelling Approach

In this modelling, the aquifer system was simplified and idealized from the real world conditions, however, the accuracy of a model is dependent upon the level of understanding of the system the model is to represent.

Table 1. Estimated daily pumping rates in wellfields and dewatering pit

Well	Estimated daily pumping rates (m^3/day)
PWM-1	6450
Well A (new)	0
Well B (new)	0
PWM-2 (JBA-1)	2640
PWM-3	4440
PWM-4	4440
PWS-1	1200
TW-4	1600
Kajibumi WF 1	0
Kajibumi WF 2	0
Imuda Tin Mine	24884
Total	45654

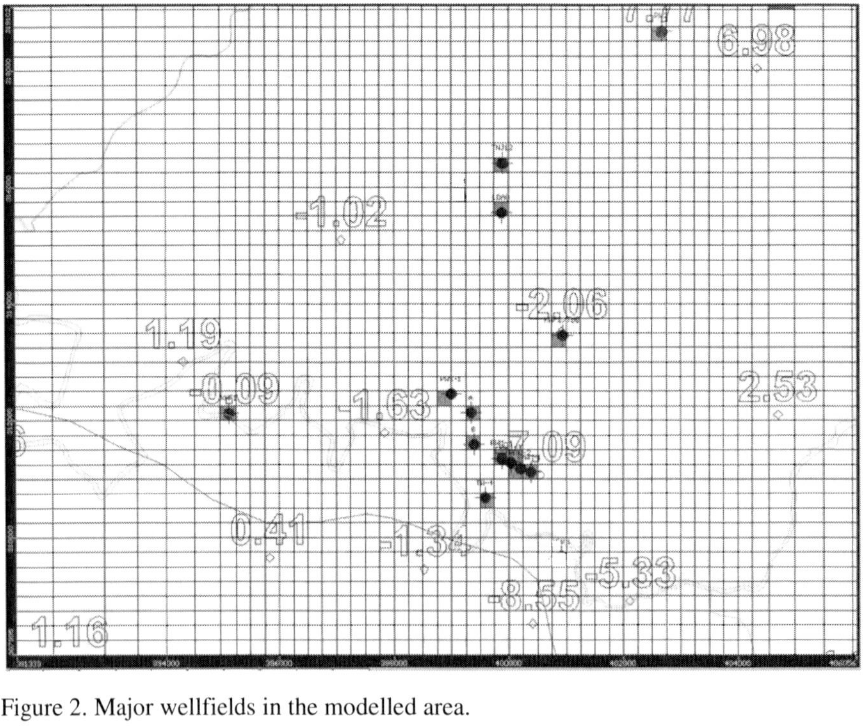

Figure 2. Major wellfields in the modelled area.

According to Anderson and Woessner (1992) groundwater flow is assumed to be strictly horizontal through the aquifers and strictly vertical through confining beds. Suratman (1997) assumed that the aquifers are continuous in his modelling work in Kota Bharu, Kelantan. Thus, a complete site investigations carried out previously and during the current study are required to accurately conceptualizes the site hydrogeology, which are necessary precursors to a successful modelling.

The modelling works involved a number of works:

- i Verification and representation of the real system in the modelled area. Realistic representation of the area was prepared for the model to give significant understanding of the real system. Various layers of the aquifers in the area were included in the model.
- ii Various boundary conditions and parameters were considered in the model and specified in order to represent the physical conditions of the area as closely as possible in the model. Certain model boundaries were generalized as certain aspects of data are limited.
- iii Development of the conceptual model, the hydrostratigraphic units and various other field parameters (such as transmissivity and storage coefficient of the aquifer) and system boundaries were assembled together with hydrologic stresses.
- iv Running of flow simulation and to measure the responses of the groundwater to the pumping stresses. Simulation of various boundary conditions including constant head, no flow and specified head, and were also represented.
- v Completion of the model. The completed model is required to be able to synthesize the data and translate the result into responses to pumping stresses and behaviour of the hydrologic system from which the results could be used for planning and management of the groundwater resources in the Study area.

The simulation could be used as a tool for planning of the sustainable groundwater resources development in the area.

The model should enable the authorities to formulate optimal management strategies leading to a development and protection of the water resources of the area.

Basically, the applied Visual MODFLOW model is a physical based modelling tool, in which most of the data required is for the model set up and input data that can be measured in the field.

The procedure of the actual modelling work involves three different steps:

1 Model set up - geometrical representation of the aquifer and physical characteristics of the system
2 Model calibration - simulation runs are carried out and model results are compared with measured data for a certain time period, and
3 Model validation - comparable measured data with modelled data

The Langat basin is a typically coastal confined aquifer with geologically defined boundaries and recharge areas. The aquifer is replenished from precipitation. Recharge areas are distributed according to geological conditions (e.g. thin or absent aquitard, bedrock outcrop). The aquifer recharge is simulated as an aerially distributed recharge to the groundwater system. The recharge is restricted to the uppermost layer and is allowed to pass through inactive cells. Aquifers can be also replenished (or drained) to some small degree from rivers (stretches with permeable riverbed). Replenishment of an aquifer can also happen also by seawater intrusion, if too much water is pumped out from a coastal aquifer. Surface runoff is simulated by shallow drains, which remove water from the first layer at a rate proportional to the difference between the head in the aquifer and some fixed head or elevation. The drain has no effect if the head in the aquifer falls below the fixed elevation of the drain.

Annual average rainfall in the area northwards from Lower Langat was estimated as 2238 mm/year and evaporation, 1284 mm/year. Thus, at least 954 mm/year is used as recharge, which infiltratesing into the cells with drains.

Model Set up

Model Grid - Discretization

The regional groundwater modelling work is concentrated in the alluvial part of the Langat Valley and the domain was selected according to natural hydrogeological boundaries, over an area with x coordinates 378000-418000 m, and y coordinates 285000-330000 m (area 40 x 45 km) as shown by Figure 3. Inside the modelling domain, a rectangular model grid with different length and width of cells has been set-up (37,240 cells). The model has a 70 x 76 cell grid in a horizontal direction; with a minimum cell size 250x250 m in the wellfield area at Megasteel, Kajibumi WF 1 and 2, Labohan Dagang, Tg. Duabelas, and Paya Indah area. The maximum cell size is 2,000 x 2,000 m.

Generally, the grid spacing is finer where the hydraulic gradient and flux are subject to greater change (such as areas near pumping wells) as shown by Figures 4a and b.

Based on comprehensive evaluation of geological logs, results of soil investigation and evaluation of pumping tests were carried out previously and from the current study. The model has been set up with seven layers in a vertical direction (Figure 5).

Aquifer Parameters

Major inputs for the model are:

- Aquifer Geometry
- Aquifer Properties
- Recharge/Discharge Mechanisms
- Boundary Conditions
- Initial Heads/Concentrations

In details, the aquifer parameters for the groundwater flow model of the Study area are listed as follows:

- hydraulic conductivity in all the active cells in x, y, z direction;
- geological properties of an aquifer in all the active cells in x, y, z direction;
- aquifer storage parameters in cells;
- aquifer active porosities; and
- parameters of boundary conditions (include geometry, water levels, parameters of leakage for general head boundary, drains, rivers, etc.).

Some assumptions and justifications were also made in the modelling work.

Initial and Boundary Conditions

Initial conditions represent the initial state of the considered model parameters and variables for starting of computation. From the point of practical groundwater flow modelling it means primary definition of groundwater level and piezometric head in each active model cell.

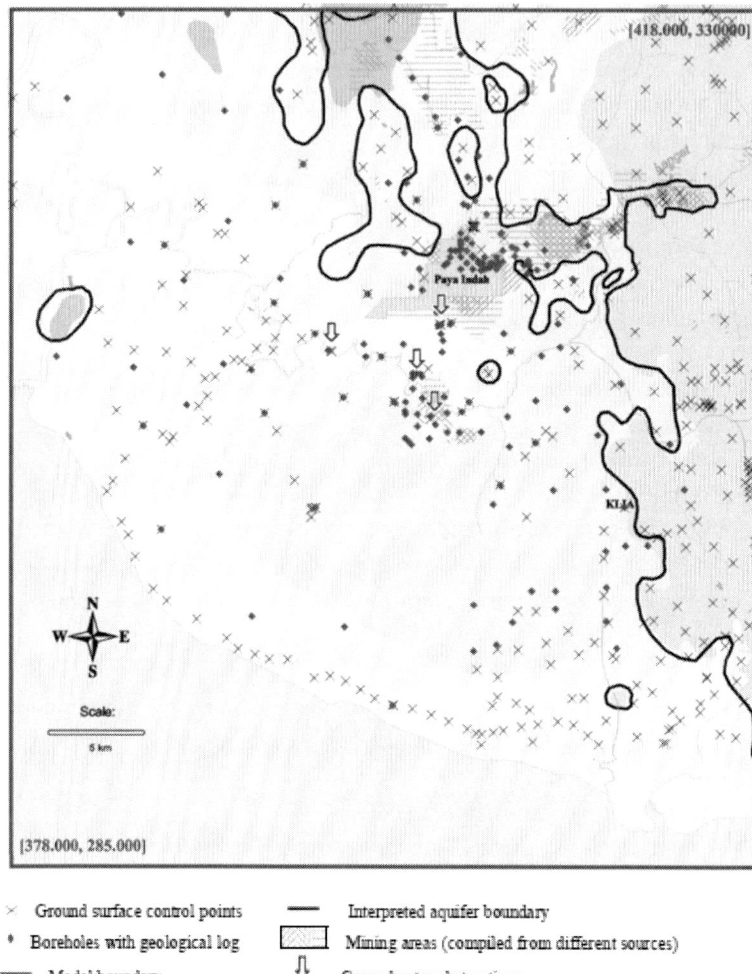

After JICA 2002.

Figure 3. Model Boundaries.

All Visual MODFLOW steady-state groundwater flow simulations (Simulation Scenario 1, 2, and 3) used the initial head values imported in the Run Module from ASCII file (file InitHead_Zdrain.TXT). The elevation of the initial head is 1 m below the ground surface.

Various boundaries were specified in order to represent the physical condition in the area.

The boundary conditions are constraints imposed on the model grid that express the nature of the physical boundaries of the aquifer being modelled. Boundary conditions have great influence on the computation of flow velocities and heads within the model area.

A no flow boundary (zero-flux, barrier) is defined as a boundary across which there is no flow of groundwater. The bottom of the model, outside of the model domain, and inactive cells are treated as a no flow boundary.

Very low permeability dredge ponds, or elevated bedrock form another type of partial barriers, which are characterized according to properties and configuration of layers. Table 2 shows the hydraulic parameters of aquifer used in the 3D model set up.

The effective recharge from rainfall is restricted to the uppermost layer and is allowed to pass through inactive cells. The value 954 mm/year was used for Simulation Scenario 1, 2, and 3 whereas in Simulation Scenario 2 the recharge value of 2400 mm/year was also run to compare the effect of higher recharge in the study area.

The River Package in the Visual MODFLOW was used to simulate the river-aquifer interaction in the study area. The river boundary in the model was represented by the Langat River as shown in Figure 4b.

Data on elevations of the river bottom and the elevation of water level were set according to values presented in Figure 6.

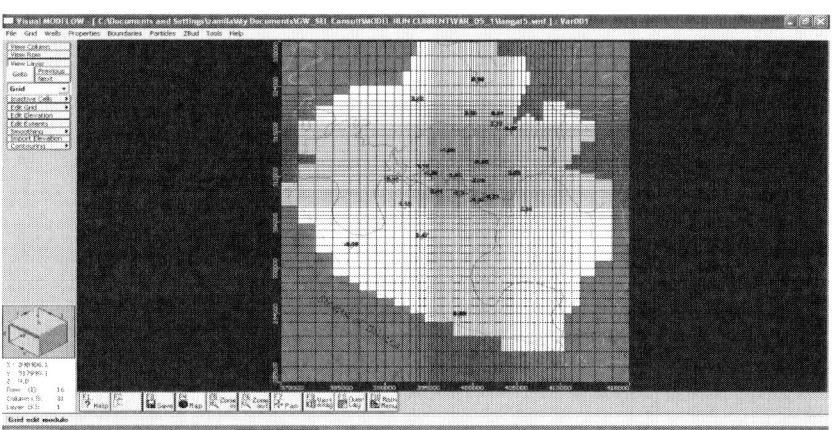

a

Figure 4. (Continued).

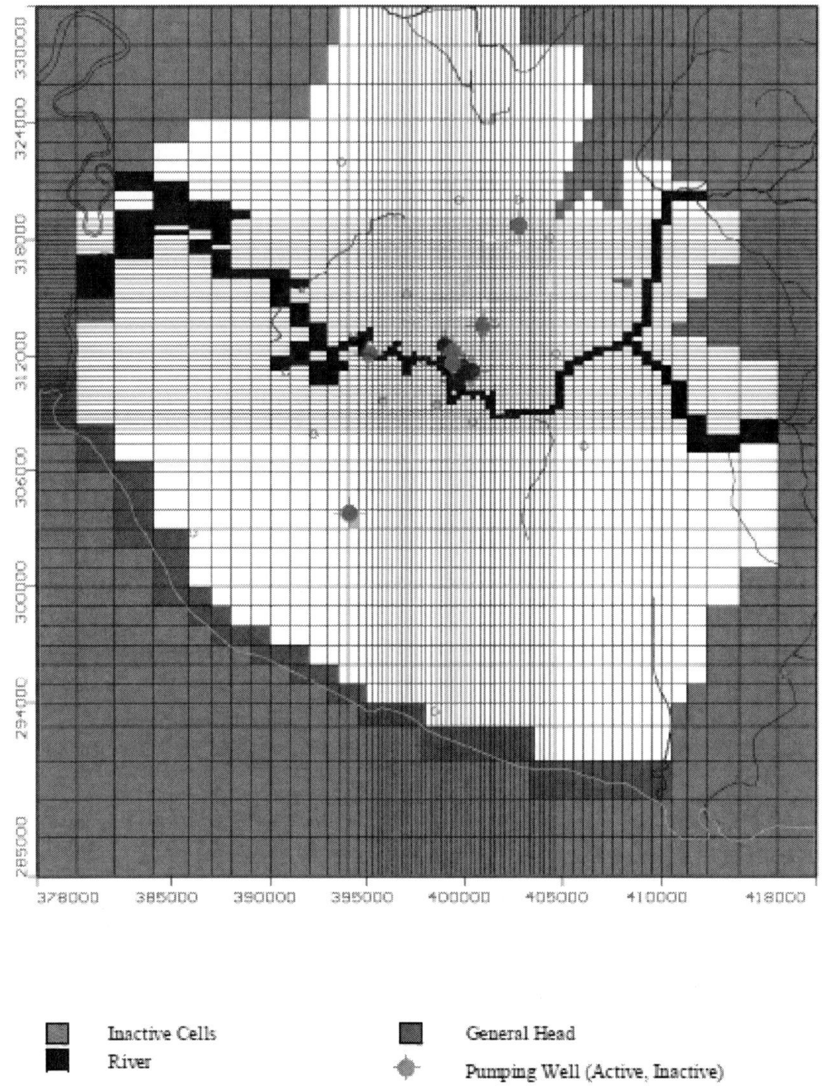

b

Figure 4. a) Model Grid and Boundaries as seen in MODFLOW interface. b) Model Grid and Boundaries.

Riverbed conductance reflects the length and width of the river in a cell. For the model, hydraulic conductivity and thickness of riverbed material was taken as $k/b=3x10^{-8}/s$.

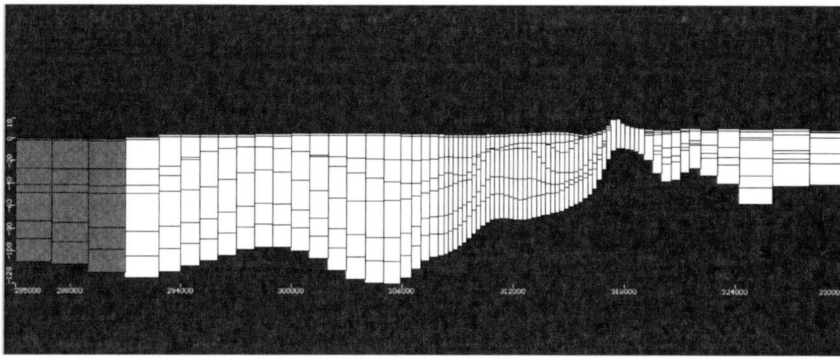

Figure 5. A model cross-section and elevation map along the centered of the Basin.

Table 2. Hydraulic parameters of aquifer used in the model set up

Layer No.	Zone	Prop. ID	Conduct. KH (m/s)	Conduct. KV (m/s)	Anisotropy K_H/K_V	S_S (/m)	S_Y (-)	Soil Description
1		18	$1.00x10^{-4}$	$1.00x10^{-5}$	10	$1.00x10^{-3}$	0.3	Peat, laterite soil
2		8	$1.20x10^{-8}$	$9.20x10^{-9}$	1.3	$6.00x10^{-6}$	0.01	Silty clay
3	A	21	$6.00x10^{-5}$	$6.00x10^{-6}$	10	$3.70x10^{-5}$	0.15	Clayey sand with gravel
	B	1	$1.00x10^{-4}$	$1.00x10^{-5}$	10	$6.00x10^{-6}$	0.15	
	C	28	$1.6x10^{-4}$	$4.00x10^{-6}$	40	$6.00x10^{-6}$	0.15	
4		9	$1.00x10^{-8}$	$7.00x10^{-9}$	1.4	$6.00x10^{-6}$	0.01	Clayey silt
5	A	23	$4.00x10^{-4}$	$9.60x10^{-6}$	42	$4.50x10^{-5}$	0.2	Silty sand and gravel, sandy clay
	B	12	$1.20x10^{-3}$	$4.80x10^{-5}$	25	$3.00x10^{-5}$	0.2	
	C	25	$1.20x10^{-3}$	$2.40x10^{-5}$	50	$4.10x10^{-5}$	0.18	
6	A	38	$1.00x10^{-4}$	$5.00x10^{-6}$	20	$4.50x10^{-5}$	0.2	Silty sand and gravel
	B	15	$2.00x10^{-4}$	$1.00x10^{-5}$	20	$3.00x10^{-5}$	0.2	
	C	40	$8.00x10^{-4}$	$3.20x10^{-5}$	25	$4.50x10^{-5}$	0.2	
7	A	20	$3.20x10^{-6}$	$3.20x10^{-6}$	1	$6.00x10^{-6}$	0.02	Bedrock
	B	24	$1.00x10^{-8}$	$1.00x10^{-8}$	1	$6.00x10^{-6}$	0.02	
Mining Ponds		2	$5.80x10^{-7}$	$5.80x10^{-7}$	1	$4.50x10^{-5}$	1	Mining areas

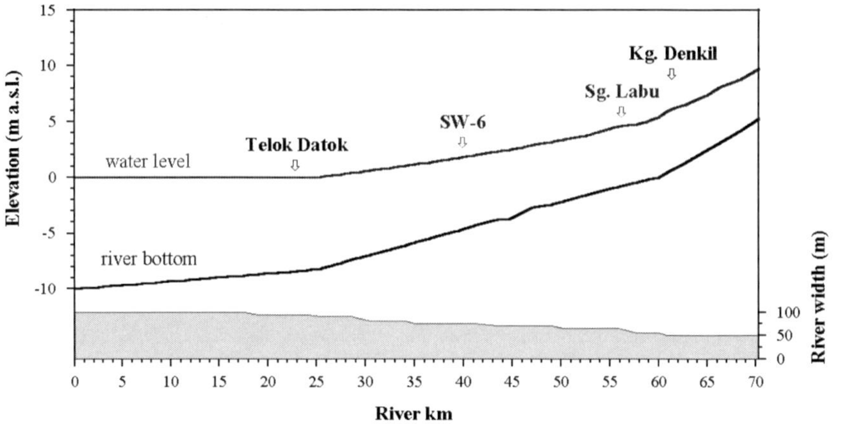

After JICA 2002.

Figure 6. Langat River properties used in the model.

Along the coast, the boundary of aquifer with the sea is represented in the model by General Head boundary as shown in Figure 4b. The water level elevation was set to 1.0 m a.s.l., to compensate for influence of density flow. The conductance represents the resistance to flow between the sea (general head boundary) and the groundwater. The values reflect the size of cells, thickness and hydraulic conductivity of sediments ($k/b=5.8x10^{-7}/s$).

The Imuda tin mine is simulated with 60 drain cells with conductance 250 m^2/d ($k/b=4.63x10^{-8}/s$). The elevation of the drain was set at -35 m above sea level. Former mining ponds around Paya Indah and the Imuda tin mine are simulated using hydraulic parameters listed in Table 3. A no-flow boundary has been assumed in the northern and western part of the model where aquifers are bounded by granitic and meta-sedimentary rocks. A value of 20% porosity has been adopted for the entire layers of the model. The location of modelled pumping wells is shown in Figure 4b; the assigned pumping rates are listed in Table 4.

Model Calibration and Verification

A mManual trial-and-error method was used with adjustments of hydraulic conductivities (model zones of homogeneous hydraulic conductivities), adjustments of leakage across a confining clayey layer, and designation of boundary conditions (river, drains, general head boundary). The model input parameters were adjusted during the calibration process to achieve a desired degree of correspondence between the model simulation and the data describing the present groundwater flow system.

Various modelling runs and calibration with some modification and optimization of initial model set up was done, to achieve convergence and to speed up the modelling computation. The most stable solver, that is Slice-Successive Over relaxation Package (SOR), was applied to ensure concrete model set up. All Simulations were performed using 0.01 m head change criterion for convergence, with acceleration factor 1.

The maximum limit on the number of iterations was set to 1000.

MODELLING RESULTS

Steady-State Simulations

The overall modelling objective is to establish a reliable groundwater model for the Langat Basin, which enables the authorities to formulate optimal management strategies leading to a sustainable development and protection of groundwater. The constructed 3D groundwater flow model represents a technical scientific based management tool and is ready for practical current and future use.

The definition of modelling scenarios is based on variation of groundwater abstraction from existing wellfields and dewatering pits; and change of groundwater recharge from rainfall as well as future pumping. Simulation Scenario 1, 2, and 3 represent steady-state flow simulation. Table 3 provides an overview of simulated pumping rates that characterize each modelling scenario. In brief the characterization of scenarios are:

- Scenario 1 – present conditions (during field investigation in the Study), with abstraction of groundwater from existing wellfields and dewatering pits (total abstraction 45,654 m^3/day).
- Scenario 2 – near future conditions, with increased abstraction of groundwater, after putting Kajibumi WF 1 and Kajibumi WF 2 into operation (total abstraction 92,251 m^3/day).
- Scenario 3 – future conditions, with increased abstraction of groundwater in the Megasteel area, with Kajibumi WF 1 and Kajibumi WF 2 in operation as well other additional wells developed in the area such Labohan Dagang and Tg. Duabelas (total abstraction 102,251 m^3/day).

Table 3. Simulated daily pumping rates in wellfield and dewatering pit (values in m^3/day)

Well	Scenario 1	Scenario 2	Scenario 3
PWM-1	6,450	13,730	13,730
Well A (new)	0	13,440	13,440
Well B (new)	0	13,730	13,730
PWM-2 (JBA-1)	2,640	2,640	2,640
PWM-3	4,440	4,440	4,440
PWM-4	4,440	4,440	4,440
PWS-1	1,200	1,200	1,200
TW-4	1,600	1,600	1,600
Kajibumi WF 1	0	9,200	9,200
Kajibumi WF 2	0	6,900	6,900
Additional new wells (Labohan Dagang (LDAG) and Tg. Duabelas (TNJ12))	0	0	10,000
Total well pumping	20,770	71,320	81,320
Imuda Tin Mine	24,884	20,931	20,931
Grand Total	45,654	92,251	102,251

Modelling Results

Output of the modelling are:

- Heads and Concentrations in each cell
- Contour maps (head equipotential, drawdown concentrations)
- Cross-sections
- Fluxes and velocities

The layout of the wells carried out in the modelling work is shown by Figure 6 in which the Scenario 2 modelling was done.

If the rainfall is considered to be 2,400 mm/year in the modelling, the mass balance solution is given in Table 4. In the modelling, the rainfall has also been reduced to see the effect of less rainfall recharging the area and the mass balance solution is given in Table 5.

In the modelling work, Layers 5 and 6 are the most important layers as they represent the most productive aquifers in the area.

Table 4. The mass balance solution when rainfall is 2400 mm/year

- **IN:**
- Storage = 0.000000
- Constant Head = 0.000000
- Wells = 0.000000
- Drains = 0.000000
- Recharge = 2971251.000000
- ET = 0.000000
- River Leakage = 1259.028600
- Head Dep Bounds = 17198.441400

- **Total In = 2989708.500000**

- **OUT:**
- Storage = 0.000000
- Constant Head = 0.000000
- Wells = 71320.203100
- Drains = 2423251.250000
- Recharge = 0.000000
- ET = 0.000000
- River Leakage = 313579.093800
- Head Dep Bounds = 181809.156200

- **Total Out = 2989959.750000**

- IN - OUT = -251.250000
- % Discrepancy = -0.010000

Table 5. The mass balance solution when rainfall is 954 mm/year

- IN:
- Storage = 0.000000
- Constant Head = 0.000000
- Wells = 0.000000
- Drains = 0.000000
- Recharge = 2739190.250000
- ET = 0.000000
- River Leakage = 2328.214400
- Head Dep Bounds = 21136.699200

- Total In = 2762655.250000

- OUT:
- Storage = 0.000000
- Constant Head = 0.000000
- Wells = 71320.203100
- Drains = 2387224.500000
- Recharge = 0.000000
- ET = 0.000000
- River Leakage = 122924.343800
- Head Dep Bounds = 181505.531200

- Total Out = 2762974.500000

- IN - OUT = -319.250000
- % Discrepancy = -0.010000

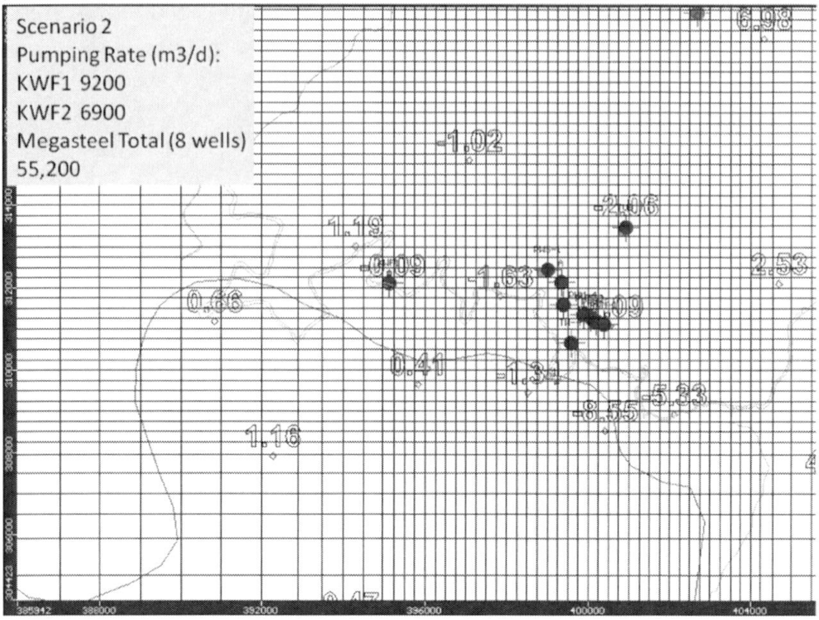

Figure 7. Abstraction wells in Scenario 2 modelling (well pumping 71,320 m^3/day; total abstraction 92,251 m^3/day).

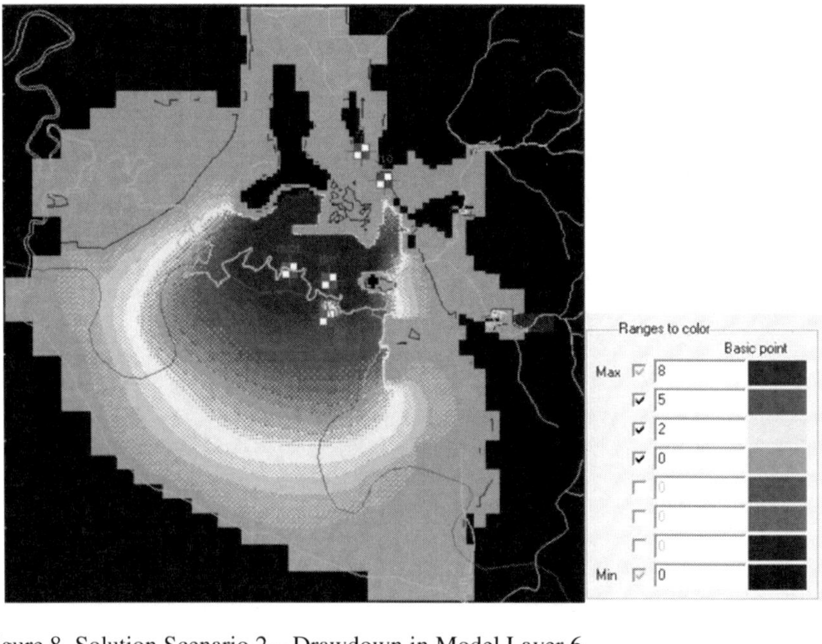

Figure 8. Solution Scenario 2 – Drawdown in Model Layer 6.

The increased abstraction in the Scenario 2 has not affected the drawdown in this layer significantly as shown by the Figure 8.

Considering the increase in pumping capacity of the wells to a total of 81,349 m^3/day with additional new wells in the Labohan Dagang and Tg. Duabelas, which could be producing in the near future, and the mass balance solution is given in Table 6 was reproduced from the modelling in Scenario 3. It could be concluded that the amount of groundwater available for abstraction is significant and able to supplement the current and future needs of the area. From the various scenarios of the modelling works in which the results are shown in Table 4, 5 and 6, significant amounts of groundwater could be abstracted as shown by the modelling work for supplying the demand of public water supply in the area. The drawdown in the most productive layers of Layer 5 and 6 in the area is shown in the Figures 9 and 10. The concentration of the drawdown is only limited within the wellfields. In The Scenario 3 of the modelling, where total abstraction of groundwater is at 102,251 m^3/day, the abstraction is still optimistically sustainable. Recharge values to the aquifers as shown by the various mass balance solution of the model shows that the amount is impressive. Recharge is in which the renewability of the groundwater is renewable. Considering taking 5% of the recharge, the available groundwater to be tapped sustainably is at 130,000 m^3/day.

Table 6. The mass balance solution with increased amount of abstraction

- IN:
- Storage = 0.000000
- Constant Head = 0.000000
- Wells = 0.000000
- Drains = 0.000000
- Recharge = 2586086.250000
- ET = 0.000000
- River Leakage = 9722.364300
- Head Dep Bounds = 26641.828100

- Total In = 2622450.500000

- OUT:
- Storage = 0.000000
- Constant Head = 0.000000
- Wells = 81349.953100
- Drains = 2361818.750000
- Recharge = 0.000000
- ET = 0.000000
- River Leakage = 1720.581100
- Head Dep Bounds = 181256.093800

- Total Out = 2626145.250000

- IN - OUT = -3694.750000
- % Discrepancy = -0.140000

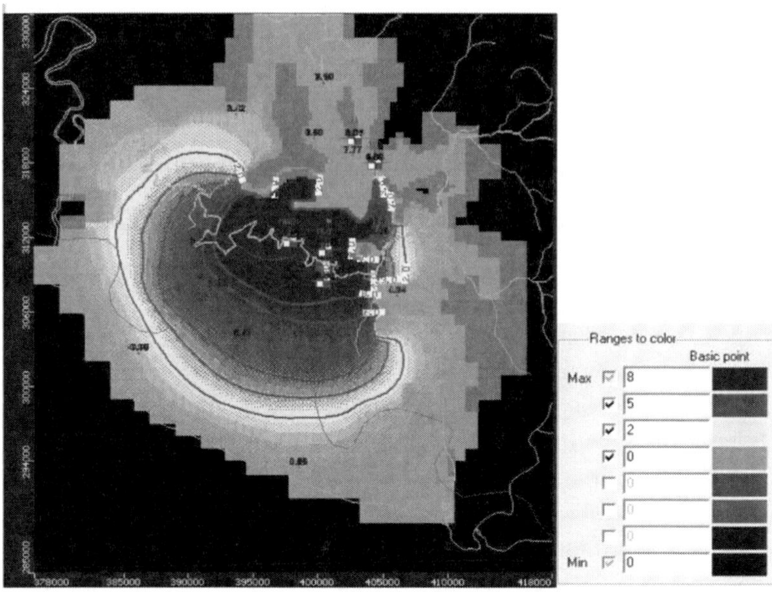

Figure 9. Solution Scenario 3 – Drawdown in Model Layer 5.

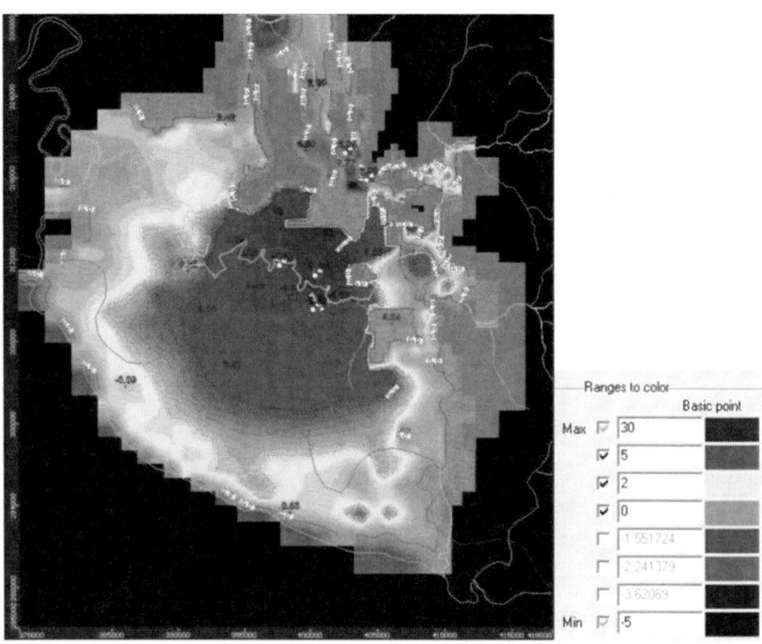

Figure 10. Solution Scenario 3 – Drawdown in Model Layer 6.

CONCLUSION

a Groundwater in the Langat Basin has been able to supplement water demands in the area especially for the industrial use.

b This study will enable JMG to recommend the viability of groundwater resources and the sustainable amounts of groundwater to the water authority.

c Groundwater could be utilized for various demands such as public water supply, industries and agriculture. The drawdown in the most productive layers of Layer 5 and 6 in the area is shown in the Figures 9 and 10. In the Scenario 3 of the modelling where total abstraction of groundwater is at 102,251 m^3/day, the abstraction is still optimistically sustainable.

d Considering the recharge in which the groundwater is renewable, taking 5% of the recharge, the available groundwater to be tapped sustainably is 130,000 m^3/day.

e It is very much recommended that the monitoring of the resources is carried out effectively to ensure that the groundwater will remain sustainable.

Reviewed by Professor Paul L. Younger, Rankine Chair of Engineering and Professor of Energy Engineering, School of Engineering, Room 623, James Watt Building (South), University of Glasgow, Glasgow G12 8QQ, Scotland, E-mail: paul.younger@glasgow.ac.uk and Dr. Ismail Yusoff, Univ. Malaya, Kuala Lumpur drismail.yusof@gmail.com

REFERENCES

Abdul Rashid, B. and Mohammed Hatta, A. K. (1998). Potensi sumber air tanah di kawasan Telok Datuk-Olak Lempit, Daerah Kuala Langat, Selangor. *Laporan Jabatan Kajibumi No. Hidro 01/98.*

Abdul Rashid, B. and Mohammed Hatta, A. K. (2000), Groundwater Development in Telok Datuk-Olak Lempit area, Kuala Langat, Selangor. *IEAG Conference 2000, Bangi, Selangor.*

Anderson, M. P. and Woessner, W. W. (1992). *Applied Groundwater Modelling: Simulation of Flow and Advective Transport.* Academic Press, London, 381 p.

Aros, M. A. (2009). Groundwater Development in Selangor and Impact to Environment. Groundwater Colloquium. *Proc. Groundwater Colloquium 2009: Groundwater Management in Malaysia - Status and Challenges,* Putrajaya, Malaysia from 25-26 March 2009.

Awadalla, S., Noor, I. M. and Ibrahim, I. (1989). Comparative Model Study for Groundwater Resources Evaluation in Kelantan, Malaysia; Part 1: Model Evaluation. In: *International Workshop on Appropriate Methodologies for Development and Management of Groundwater Resources in Developing Countries*, (Feb. 28-Mar. 4, 1989, Hyderabad, India), Gupta, C. P. (ed.), Vol. II, p. 619-627.

Drillco Technologies (M) Sdn. Bhd. (2005). *Environmental Assessment of groundwater abstraction study for Megasteel Sdn. Bhd.*

Guiguer, N. and Franz, T. (1997). *Visual MODFLOW.* Waterloo Hydrogeologic Software, Canada.

Hussin, A. (2001). General Geology of the Langat Basin, *The 2nd workshop on the study on the sustainable groundwater resources and environmental management for the Langat Basin,* Malaysia, 20th March 2001, Kuala Lumpur.

Ibrahim, A. N., Harith, Z. Z. T. and Nawawi, M. M. (2003b). Resistivity Imaging and borehole investigation of the Banting Area Aquifer, Selangor, Malaysia, *Journal of Environmental Hydrology*, V. 2, Paper 10.

Ibrahim, A. N., Harith, Z. Z. T. and Nawawi, M. M. (2003a). Using integrated 2-D electrical imaging and seismic refraction methods for study of groundwater aquifer in Brookland Plantation, Selangor, Malaysia, *Proceeding of the 6th SEGJ International Symposium - Imaging Technology*, Tokyo, Japan, 398-404.

Ibrahim, A. N., Nawawi, M. M., Harith, Z. Z. T., Ayub, M. S., and Alphonse, A. (2001). Ground water aquifer mapping using 2.5-D Electrical Imaging technique in Brookland Plantation, Selangor, Malaysia, *Prosiding Seminar Sehari Kemajuan Terkini Kaedah Geofizik Kejuruteraan,* Geological Society of Malaysia, 4.1-4.7.

JICA (2002). *The Study on the Sustainable Groundwater Resource and Environmental Management for the Langat Basin in Malaysia.* Department of Minerals and Geoscience Malaysia, Report.

JMG. (2005). Potensi Pembangunan Sumber Air Tanah di Negeri Selangor. *Report No. JMG.SWP.GS.2/2005.* Jabatan Mineral dan Geosains Selangor, 13 p.

McDonald, M. G. and Harbaugh, A. W. (1988). A Modular Three-Dimensional Finite-Difference Ground-water Flow Model, *Technique of Water-Resources Investigation Book 6, Chapter A1.* US Geological Survey, 576 p.

Mohamed Hatta, A. K. and Abdul Rashid, B. (1999). Development of a large groundwater basin in Kuala Langat, Selangor. *Persidangan Geosains Semenanjung 1999,* Hotel Heritage, Ipoh 23-25 November 1999.

National Economic Action Council (1998). *National Economic Recovery Plan: Agenda for Action.* National Economic Action Council, Economic Planning Unit, Prime Minister's Dept.

Noor, I. M. (1980). Prefeasibility study of potential groundwater development in Kelantan, Malaysia. *Ph.D. Thesis*, University of Birmingham, United Kingdom, (unpublished), 427 p.

Suratman, S. (1997). Groundwater Protection in North Kelantan, Malaysia: An Integrated Mapping Approach Using Modelling and GIS. *Ph.D. Thesis,* University of Newcastle Upon Tyne, United Kingdom.

Suratman, S. (1999). Groundwater flow modelling of the Teluk Datuk-Olak Lempit, Kuala Langat, Selangor. Jabatan Penyiasatan Kajibumi Malaysia, *Report no. GB 1/99, 20 p.*

Suratman, S. (2005). The study on sustainable groundwater resources and environmental management for the Langat Basin in Malaysia. In: *Groundwater Intensive Use: IAH Selected Papers on Hydrogeology 7,* Sahuquillo, A., J. Capilla, J., Cortina, L. M. (Editor), and Sanchez-Vila, X. (Eds.). Balkema Pub., Leidin, 341-350.

Umar, H., Abdul Rahim, S., Abdul Rahman, Y., Rashid, B., and Abdul Ghani, R. (2000). Penggunaan Teknik Geofizik dan Geokimia untuk mencirikan Struktur Subpermukaan dan Akuifer di Sekitar Olak Lempit, Selangor, *Proceeding Annual Geological Conference 2000*, Geological Society of Malaysia, 381-386.

In: Groundwater
Editor: Segun Michael Adelana
ISBN: 978-1-63321-759-1

Chapter 7

FLOW, RECHARGE AND MIXING PROCESSES IN THE WERRIBEE BASIN (AUSTRALIA) USING NATURAL ENVIRONMENTAL ISOTOPE GEOCHEMISTRY: IMPLICATIONS FOR WATER RESOURCES MANAGEMENT

S. Michael Adelana[1]*, Graeme Allinson[2,3], Matt Kitching[2], Hiroyuki Ii[4], Scott Salzman[5], Elissa McNamara[6], Masonobu Taniguchi[5], George Croatto[2] and Bruce Shelley[2]

[1]Agriculture Research, Department of Environment & Primary Industries, DEPI Bendigo Centre, Victoria, Australia
[2]Agriculture Research, Department of Environment & Primary Industries, DEPI Macleod, Victoria, Australia
[3]Centre for Aquatic Pollution Identification and Management, % Department of Chemistry, The University of Melbourne, Parkville, Australia
[4]Faculty of Systems Engineering, Wakayama University, Japan
[5]School of Information Systems, Deakin University, Warrnambool, Victoria, Australia
[6]Southern Rural Water, Victoria, Australia

* Corresponding author: Michael.Adelana@depi.vic.gov.au.

Abstract

Chloride (Cl) and stable isotope ($\delta^{18}O$, δD) compositions in rainfall, river water, shallow and deeper groundwaters have been used to investigate the recharge, flow and mixing processes through the aquifer system within the Werribee basin, southwest of Melbourne (Australia). Total Dissolved Solids (TDS) contents and nitrate concentrations are generally lower at the upper and mid-catchments compared to the lower catchment. The Cl^- and TDS, together with $\delta^{18}O/Cl^-$ and Cl^-/Br^- ratios imply that evaporation (e.g. from shallow watertable) and transpiration (e.g. from the all-year-round cropping) during recharge represent the main processes controlling groundwater composition, in addition to mineral weathering. Measured groundwater $\delta^{18}O$ and δD values scatter around the local meteoric water line indicating that groundwater is primarily of meteoric origin... The shallow groundwater, particularly at the delta (delta aquifer) must have been derived from a combination of sources: the sea, recycled water and river delta water from surface irrigation and upstream groundwater from upper catchment area. There is evidence of vertical mixing in the basin based on the TDS distribution and elevated NO_3^- concentrations up to 45 m depth below ground surface. However, TDS and NO_3^- contents generally decrease and homogenizes with depth. These are consistent with gradual vertical mixing (calculated mixing ratios) from between saline shallow borehole delta and lower salinity deeper groundwater in the Brighton Group aquifer. Using isotope geochemistry to identify where potential surface contamination may occur and/or to confirm seawater intrusion is an important contribution to protecting valuable groundwater resources within the metropolitan city of Melbourne.

Keywords: Coastal aquifer - hydrological processes - Groundwater flow - Chloride - isotopes – Salinization

Introduction

The water environment is under increasing pressures from changes in land-use, increased demand for water supplies, and climate variability. Studies have shown the gap between freshwater supply and demand will widen during the coming century as a result of these changes and as a consequence of increasing consumption of water by a growing human population (Bates et al., 2008). In many regions of the world, the amount and quality of water available to meet this growing human needs are already limited (CCSP, 2008; Pittock,

2011). These pressures create challenges for managing catchments in many parts of the world. In Australia, similar pressures have been imposed on local groundwater and surface water resources.

The combined effects of increasing water demand and climatic characteristics of Australia, (with recurrent droughts), during which regulated water reserves come under greater pressure, has created significant challenges for sustainable management of water resources.

The management of catchments containing important groundwater and surface water resources has witnessed several research efforts worldwide, including the use of environmental isotopes, e.g., in Europe and the Americas (Dincer et al., 1970; Mook et al., 1974; Rodhe, 1984; Christopherson et al., 1985; Kennedy et al., 1986), in Asia (Sklash et al., 1976; Sklash and Farvolden, 1982; Saha et al., 2013) and in Australia (Bennetts et al., 2006; Banks et al., 2011; Cartwright et al., 2012). Due to the complexity of many hydrogeological systems and associated flow processes, environmental isotopes can serve as a valuable tool to offer additional insights into ground and surface water processes (e.g. Clark & Fritz, 1997; Herczeg et al., 2001).

The stable oxygen and hydrogen isotopes of water (usually designated as $\delta^{18}O$ and $\delta^{2}H$ or δD) have found wide application for determining recharge mechanisms (Harrington et al., 2002), origins of salinity (e.g. Herczeg et al., 2001) as well as surface water-groundwater interactions (McCarthy et al., 1992). These varied applications rely on the fact that water retains its stable isotopic signature unless diluted or mixed with waters of a different isotopic composition, or unless fractionated by evaporation or isotopic exchange. Conservation of stable isotopic composition of water permits the identification of groundwater recharge areas and flow within the system, which has been used successfully in many semi-arid to arid regions (examples in Clark & Fritz, 1997). Natural variations in stable isotopic composition of components of the catchment water balance can, under favourable conditions, be used to identify such important processes as flow patterns that lead to the generation of streamflow from a catchment and their pathways (Turner et al., 1987). Both groundwater dynamics and groundwater contamination can be influenced by hydraulic interconnections between aquifers. Environmental (stable) isotopes can be used to investigate such interconnections, based on contrasting compositions. Intense exploitation of groundwater can induce leakage from overlying and underlying aquifers. Stable isotope data, therefore, can be used to estimate the flow of groundwater from adjacent aquifers.

Moreover, isotope geochemistry is very relevant in pollution studies. For example, nitrate-nitrogen isotopes can be used as an in-situ tracer of effluent in groundwater; and because it is negatively charged, nitrate in groundwater can potentially be transported across long distances under amenable conditions. As such, groundwater nitrate could ultimately aid in delineating plumes and protecting groundwater.

This study, based on the observed isotopic variations of precipitation, groundwater and surface water resources, attempts to establish the different components of the groundwater system within the catchment as well as flow mechanisms in the Werribee deltaic and basalt aquifers.

The specific objectives are: (1) to use stable isotopes to develop the Local Meteoric Water Line (LMWL) for the Werribee Basin and calculate local evaporation slope; (2) to characterize the isotopic composition of river water and groundwater and explore the recharge processes; (3) to understand connectivity of surface and ground water resources in the irrigation districts, 4) to understand the origins and evolution of salinity in the groundwater and surface water systems

THE WERRIBEE RIVER CATCHMENT

Catchment Description and History of Development

The catchment of the Werribee River (also known as the Werribee basin) covers an area of approximately 2,348 km^2, and is located in central Victoria 20 km to the west and south west of Melbourne (Figure 1). Large areas of the eastern part of the catchment are flat with poor drainage characteristics.

Table 1(a). Seasonal temperatures and average rainfall totals for the seasons within the study period

Season	Average Maximum Temperatures (°C)	Average Minimum Temperatures (°C)	Average Rainfall Totals (mm)
Summer (December–February): warm to hot	26.6	15.1	148
Autumn (March–May): mild	21.4	11.5	122
Winter (June–August): cool to brisk	15.1	7.0	111
Spring (September–November): cool to mild	21.3	10.2	160

(Data source: Bureau of Meteorology, Climate statistics for Melbourne (2006-2011)

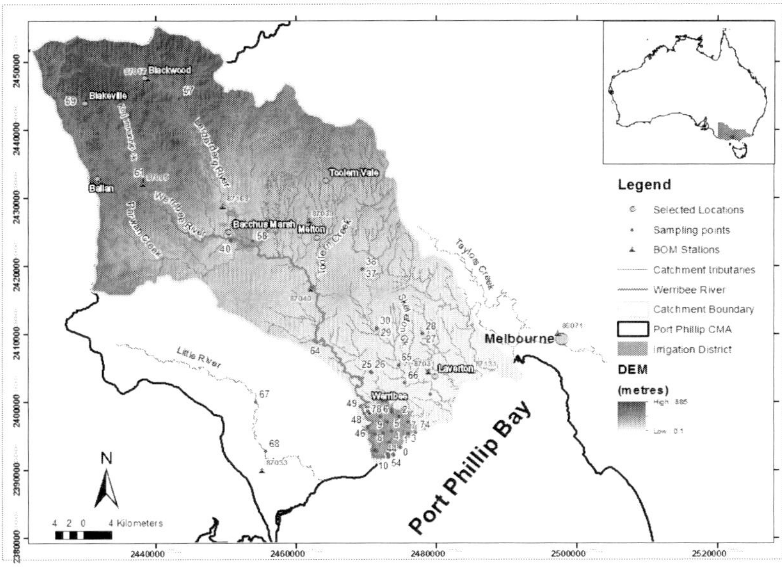

Figure 1. Location of the Werribee River catchment showing irrigation districts and sampling points.

Table 1(b). The capacity and surface area at full supply level for the major reservoirs located within the Werribee River catchment (After Lowe et al., 2009)

Reservoir Name	Capacity (ML)		Surface area at full supply level (ha)	*Average surface area (ha)	**Evaporation (ML/yr)
	Min	Max			
Melton Reservoir	14	360	245	148	1214
Merrimu Reservoir	32	516	310	103	1340
Pykes Creek Reservoir	22	119	203	76	990

*average surface area in 2005/2006.

**estimated evaporation from reservoir during 2005/2006 (data from Lowe et al., 2009)

The Werribee River flows through a landscape that varies from hills and gorges to flat plains, and flows directly into Port Phillip Bay; some of the catchment is now part of metropolitan Melbourne (SRW 2009). The region has a wet winter (June-August) and warm to hot, dry summers (December-February). For instance, the long-term (1950-2011) median rainfall and mean Class A pan- evaporation at Werribee/Laverton are 538 and 1268 mm/yr, respectively (Table 1(a)). Water impounded from both the Werribee and Lerderderg River systems is stored in three main reservoirs: Pykes Creek, Merrimu (on Pyrite Creek in Figure 1) and Melton Reservoirs (Table 1(b)).

While the Werribee River is used to meet some urban water demands, its resources are predominantly used for agriculture, including the irrigation of market gardens, perennial pasture and orchards. Two main irrigation districts are enclosed within the Werribee catchment (Figure 1): the Bacchus Marsh Irrigation District (BMID) and the Werribee Irrigation District (WID); in addition, the Lerderderg Irrigation Trust also had a long history of irrigating about 810 hectares of river flats along the banks of the Lerderderg River (SRW 2009).

The WID receives its freshwater irrigation supply via the Werribee River from water stored in the reservoirs. This reticulated water has at various times been supplemented by groundwater, and stormwater diverted from runoff drains. Since 2004, recycled water taken from the WID Recycled Water Scheme has been available and used to mix with river water or groundwater to reduce salinity, although the overall contribution of each component varies annually. The history of usage has been summarized in Allinson et al. (2011) as follows:

(i) River water: (pre-1998) agricultural activity used 13-16 GL/yr, reduced to ~600 ML/yr on average during the drought years post-1998.
(ii) Groundwater: pre-1997 use, ~2 GL/yr and licensed allocations 5.1 GL/yr, although present use has exceeded this level (e.g. ~6.4 GL in 2003-2004).
(iii) Storm water: such diversions contribute ~275 ML/yr in flow into the WID.
(iv) Recycled water: since 2003, ~13.6 GL/yr (average).

Basin Morphology, Geology and Hydrogeology

The Werribee River basin is in itself part of a larger unit, the Port Phillip Sunkland (Birch, 2003). The delta lies above the Parwan Trough, which contains a sequence of sedimentary and volcanic rocks with a thickness of approximately 300 m, deposited during the Palaeogene and Neogene (Holdgate et al., 2002). The area surrounding the Werribee River comprises mainly of deltaic sediments e.g. sand, silts and clays (SKM 2005). The soils from selected soil profiles (GeoAust 2009) on the Werribee River Delta deposits consist of an thin upper layer of windblown and clayey silts (up to 2.1 m thick). A layer of residual basaltic clay (thickness up to 5.2 m) underlieays the silt. Weathered basalt is found at depths greater than 0.7 m under the silt and clay layers (Coombes et al., 2010). The soils of the WID exhibit slightly elevated levels of nutrients, pH, sodium and salinity,; which are a likely consequence of over 80 years of irrigation (with water of increasing salinities) together with extensive and intensive fertiliser usage (SRW, 2009; Wei et al., 2013).

The stratigraphy of the major hydrogeological units of the area are, with increasing depth, (1) the Werribee Delta sediments (Quaternary alluvium), and (2) the Newer Volcanics (Quaternary basalts). The Werribee Delta sediments are of variable thickness, ranging from 30 m to 160 m thick (SKM, 2004; Leonard, 2008; Holdgate et al., 2002). The Newer Volcanics, which will be hereafter referred to as basalts, are a layer of basalt covering most of the Werribee Delta (Holdgate et al., 2002),. Underneath this basalt isoverlie, in stratigraphic order a marine to non-marine layer of clay, silt and fine sand defined asof the Brighton Group (SKM, 2005). alluvial and marginal marine sediment The deepest stratigraphic element isof the Werribee Formation (SKM, 2005), a layer found more than 150 m below the ground surface, and Silurian comprised mainly of alluvial and marginal marine sedimentary deposits. BBedrock of Silurian origin underlies the Werribee Formation (Holdgate et al., 2002). The most permeable horizons are believed to be the fractured basalts and the coarser sediments immediately overlying the basalts (Hare et al., 2005).

Groundwater management in the Werribee Delta is challenging. Connections to deeper aquifers, along with tidal and river influences, complicate the hydraulics of the system. Across the WID, the uppermost layer is generally the clay, silt, sand and gravel of the delta sediments, which is considered to be the primary or principal aquifer in the area based on its permeability (SKM 2005). These sediments form a discrete groundwater flow system hydraulically connected to the underlying basalts and Brighton Group (Dahlhaus et al., 2004). In both Werribee Delta and basalt aquifers, groundwater is essentially unconfined. Recharge into the basalt aquifer is also thought to be byboth through outcrop and leakage from the overlying Lower

Werribee Delta, where basalt is not outcropping. There are is no evidences of confining beds between the upper and the lower aquifers, and the existence of intra-volcanic sands lenses enhances connectivity (Leonard, 2008) also and promotes recharge by leakage.

Bore yields generally range between 2 l/s and 10 l/s; sometimes as high as 40 l/s (SKM, 2005). Hydraulic conductivity (lateral permeability) is extremely variable, ranginge from 10^{-6} m/d to 10^{2} m/d, with clayey materials <1 m/d and sandy material 10 m/d to 15 m/d. Aquifer transmissivity was estimated to be generally less than 20 m^2/d while storativity is variable, from 0.01 to 0.1 (Dahlhaus et al., 2004). The electrical conductivity of groundwater is generally between 800 and 3,000 µS/cm but over 8,750 µS/cm in some bores. For the WID, tThere are over 200 groundwater licenses issued to extract a maximum volume of approximately 5,100 ML/year (SRW, 2009).

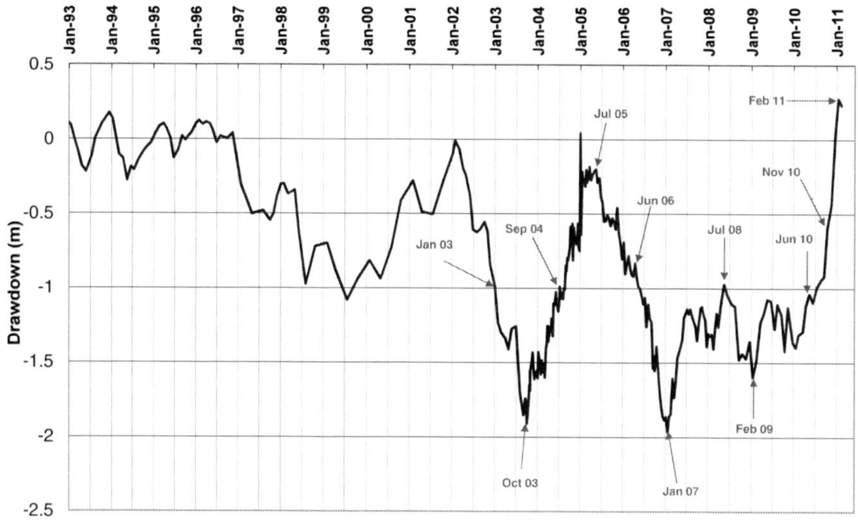

(Note: The Deutgam aquifer is a thin shallow alluvial deposits that are associated with river valleys within the Deutgam Local Government Area, otherwise declared a Water Supply Protection Area for permanent trading of water and management purposes. About 25 licensed and monitored bores have their water levels averaged from a daily timestep).

Figure 2. Average drawdown in Deutgam Water Supply Protection Aquifer (SRW 2009). Note: The Deutgam aquifer is a thin shallow alluvial deposit that is associated with river valleys within the Deutgam Local Government Area, otherwise declared a Water Supply Protection Area for permanent trading of water and management purposes. The dates indicate seasonal fluctuation peaks and lows.

Monitoring of long-term trends in groundwater heads and extensive boreis possible through hydrograph records of >20 years exist for many sites bores (VVG, 2011). These records have been used to examine water level fluctuations with net drawdown (Figure 2) in the shallow Delta delta aquifer and variations in recharge in this catchment (SRW, 2009). For example, about 25 licensed and monitored bores have their water levels averaged from a daily timestep (Fig. 2). Groundwater flow direction generally follows from the topographic highs in the upper catchments around Ballan to the lower surface elevation with respect to mean sea level near the coast. Recharge occurs across the catchment due to extensive outcrop of basalts in the Werribee Plains and along a longitudinal N-S trend across the basin; estimates are in order of 5% to 10% of rainfall, with additional increments from irrigation. Higher recharge can occur where the Werribee plain is underlain by younger volcanic rocks and scoria (Dahlhaus et al., 2004).

MATERIAL AND METHODS

Sampling Network

The Werribee River waters were classified into three zones based on catchment topographic features (Ii et al., 2012). Specifically, river delta sites were on the Werribee River less than 20 km from the coast; river medium sites were mid-catchment locations on the Werribee River 20 to 40 km from the coast; river upstream sites were on the Werribee River 50 to 60 km from the coast (Table 2a). The river water samples include those collected from tributaries of the Werribee River (the Lerderderg River, Parwan Creek, and Toolern Creek; excluding the lower stream area) and the rivers bordering the Werribee delta on its west and eastern sides (Little River, and Skeleton Creek) (Figure 1).

Sampled bores were from a collection of 25 State Observation Bore Network (SOBN) and four private bores in the WID and, two private bores adjacent to the WID in the Werribee Formation. Eight SOBN bores screened in the basalt aquifer to the north and east of the WID, and four SOBN bores in the Werribee Formation/Brighton Group to the north west of the WID were also sampled (Table 2b).

Sea, river, bores, recycled and rain waters were sampled between 2007 and 2010. Sea water was sampled at the delta coast, in Port Philip Bay and the

Werribee River estuary. Bore waters were sampled from the mid-catchment area to the delta area (Figure 1).

The Werribee River and its surrounding river waters were sampled along their courses. Recycled waters were sampled from irrigation supply channels and drains in the delta. Rain water was sampled at Werribee DPI Office. Temperature, pH, oxidation reduction potential (ORP) and electrical conductivity (EC) values were measured on site using a portable instrument (HORIBA, pH meter D-54, 55). Samples for anions and cations, filtered through 0.45 μm filters, were measured using liquid ion chromatography (DIONEX, DX-ICS, AQ-1,500 and ICS-1,600).

Within the WID agricultural runoff, stormwater and irrigation channels were sampled. For repeat samples taken in different years, and for groundwater samples taken from the three different aquifers, representative samples were selected and several well volumes pumped prior to sampling. Stable isotope samples were taken in 100-ml HDPE screw-cap bottles and sealed tightly to prevent evaporation. In order to define the LMWL, over four hundred samples of rainfall taken in Melbourne between 1962 and 1998, analysed as part of the IAEA Global Network of Isotopes in Precipitation (GNIP), and an additional 102 rainfall samples monitored for the stable isotopes of rainfall in nearby centres were included to define the LMWL. These are isotope data from long-term monitoring retrieved from the archives of WISER (IAEA/WMO, 2006). Six precipitation sampling stations (e.g. Melbourne, Camberwell, Carlton, Preston, Hampton East, Melton) were set up by the Melbourne University Network of Isotopes in Precipitation (MUNIP) to collect the bulk rain samples that generated this GNIP dataset while few rainwater samples were sampled during specific rain events at Werribee DPI Office (Table 3). Clark & Fritz (1997) also demonstrated that using a large precipitation dataset will improve the confidence level of the meteoric water line, allowing a better interpretation of groundwater data. Therefore, the long-term data is representative of long term averages of isotopic composition of precipitation within the basin. Recent monitoring of rainfall $\delta^2H/\delta^{18}O$ composition from April 2008 to May 2010 (Crosbie et al., 2012) was used to update the IAEA long-term data for the stable isotopes of precipitation in Melbourne.

Daily rainfall data collated from Bureau of Meteorology (www.bom.au) were summed into monthly values and then analyzed to characterize the rainfall pattern in the study area and seasonal variations relevant to stable isotopes studies.

Analytical Methods

The oxygen and hydrogen stable isotope ratios of water samples were measured using a mass spectrometric system (Finnigan Mat Delta Plus), followed by the CO_2-H_2O and H_2-H_2O equilibration technique using platinum catalyst (Epstein & Mayeda, 1953; O'Neil et al., 1975). All samples were analyzed for oxygen-18 and deuterium at Wakayama University, Japan. The measured oxygen isotopic ratio was calculated using equation (1).

$$\sigma\ (‰) = (R_x/R_{st}-1) \times 1000 \qquad (1)$$

where R_x is the stable isotopic ratio of x, sample and R_{st} the stable isotopic ratio of st, standard.

This conventional notation as per mil (‰) deviations from the International Atomic Energy Agency (IAEA) Standard Mean Ocean Water (SMOW)/Standard Light Antarctic Precipitation (SLAP) reference standards. Analytical precision is ± 0.2‰ for $\delta^{18}O$ and ± 2.0‰ for δD.

Table 2a. Sampled points from the Werribee River catchment area (classified according to catchment location)

Sampled water	Distance from coast	Character
River delta (Werribee River)	Less than 20 km	Delta and plain, lower catchment
River medium (Werribee River)	20 to 40 km	Plain, middle catchment
River upstream (Werribee River)	50 to 60 km	Mountain, upper catchment
Borehole delta (groundwater)	Less than 10 km	Werribee Delta, intensive irrigation, lower catchment
Borehole medium (groundwater)	10 to 20 km	Plain, agriculture, lower catchment
Borehole upstream (groundwater)	20 to 40 km	Plain, agriculture, middle catchment

Table 2b. Characteristics of the groundwater monitoring points from the sampling network and descriptive statistics of measured groundwater levels in the period 1992-2011

Site No.	Easting GDA94z55	Northing GDA94z55	Bore Depth (m)	Class	Screen interval (m)	Formation	Water table elevation (m AHD)					
							No.	Min	Max	Med	Mean	SD
G-20	2474905	2393352	13	a	7.0-12.8	Werribee Delta	360	-0.2	1.2	0.5	0.5	0.2
G-21	2474905	2393352	29	c	26.0-28.0	Delta/Brighton	360	-0.2	1.0	0.3	0.3	0.2
G-22	2474720	2397776	23	b	11.8-23.0	Basalt	361	4.9	8.2	6.6	6.6	0.8
G-23	2476029	2395408	32	b	11.0-30.0	Werribee Delta	363	2.6	5.0	4.0	3.9	0.6
G-25	2473596	2395730	14	b	8.5-14.0	Werribee Delta	376	3.3	10.1	7.2	6.9	1.3
G-26	2473596	2395730	23	b	18.0-22.0	Basalts	373	3.3	10.1	7.0	6.8	1.4
G-28	2472042	2397864	20	b	9.0-20.0	Basalts	361	6.1	11.1	8.0	8.3	1.5
G-30	2476029	2395408	55	d	31.0-55.0	Basalts-Brighton	363	5.3	6.0	5.6	5.6	0.1
G-31	2471151	2395533	26	c	20.0-26.0	Delta/Basalt?	359	0.5	2.2	1.2	1.2	0.4
G-32	2471151	2395533	78	d	29.0-40.2	Brighton	351	0.7	4.9	1.5	2.4	1.5
G-33	2473254	2391947	43	b	14.2-20.2	Werribee Delta	357	0.1	1.5	0.4	0.4	0.1
G-34	2471275	2392910	23.5	c	18.1-23.5	Delta-Basalt?	361	-0.7	3.7	2.2	2.0	1.1
G-35	2471275	2392910	31	c	23.7-31.0*	Basalts-Brighton?	361	-0.8	3.2	2.1	1.9	0.9
G-36	2473381	2394085	41	b	17.0-23.0	Werribee Delta	359	0.8	6.0	3.6	3.7	1.3

Site No.	Easting GDA94z55	Northing GDA94z55	Bore Depth (m)	Class	Screen interval (m)	Formation	Water table elevation (m AHD)					
G-37	2474905	2393352	35	b	12.0-18.0	Werribee Delta	356	-0.3	1.2	0.5	0.5	0.2
G-38	2472965	2397074	27	b	13.0-27.0	Basalts	360	5.7	12.5	9.5	9.4	1.5
G-39	2472965	2397074	10	a	2.7-8.5	Werribee Delta	359	6.0	12.8	9.8	9.6	1.5
G-02	2471347	2395328	18	b	11.9-16.3	Werribee Delta	284	2.1	6.1	3.8	3.8	1.1
G-03	2471512	2392775	27	c	19.5-25.5	Delta/Basalts?	259	-0.3	2.3	1.0	1.0	0.4
G-04	2473152	2392281	27	c	19.0-25.0	Brighton?	284	0.3	2.2	0.9	1.0	0.4
G-18	2473152	2392279	8	a	5.0-8.0	Werribee Delta	284	0.3	2.2	0.9	1.0	0.4
G-70	2472042	2397864	10.6	a	7.6-10.6	Werribee Delta	216	6.4	11.4	7.8	7.9	0.9
G-71	2471151	2395533	20.4	b	17.0-20.0	Werribee Delta	216	0.4	2.0	0.8	0.9	0.3
G-72	2471275	2392910	15	b	11.0-14.0	Werribee Delta	216	-0.9	3.6	1.3	1.3	1.0
G-73	2476029	2395408	11.5	a	7.5-10.5	Werribee Delta	216	2.6	4.8	3.5	3.5	0.5
Gs19	2473221	2392506	13.1	b								
G069	2471924	2396093	19.0	b	15-19		360	5.7	12.5	9.5	9.4	1.5
Gs05	2471602	2410834	24.0	c	6-18							
Gs06	2470731	2404400	38.0	d	24-36							
Gs00	2478104	2410045	38.0	d	32-38		249	30.2	7.9	22.3	15.9	9.6

Table 2b. (Continued)

Site No.	Easting GDA94z55	Northing GDA94z55	Bore Depth (m)	Class	Screen interval (m)	Formation	Water table elevation (m AHD)					
Gs01	2478124	2410042	29.0	c	17-29		252	30.1	8.3	21.8	16.0	9.8
Gs84	2471606	2410834	44.5	d								
Gs86	2470712	2404407	40.2	d			359	6.0	12.8	9.8	9.6	1.5
Gs28	2473034	2394732	16.0	b	11-16		284	2.1	6.1	3.8	3.8	1.1
G906	2479284	2401178	176.9	d	149.8-164.8		259	-0.3	2.3	1.0	1.0	0.4
Gs63	2450915	2423823	26.3	c	20-23		284	0.3	2.2	0.9	1.0	0.4
Gs65	2450917	2423822	12.0	b	5-8		284	0.3	2.2	0.9	1.0	0.4
Gs68	2451176	2426234	30.0	c	26-29		216	6.4	11.4	7.8	7.9	0.9
Gs69	2451180	2426234	12.7	b	8.7-11.7		216	0.4	2.0	0.8	0.9	0.3
Gs39	2470299	2398367	260.0	d	240-252		216	-0.9	3.6	1.3	1.3	1.0

Estimation of Mixing Ratios

First, the fact that Cl^- concentration of the sea was extremely high relative to other waters, Cl^- was used for determining the mixing ratio of sea water. The mixing ratio between sea and river water (recharged water) was calculated using equation (2).

$$M_{sea1} = (Cl^-_{sam}-50)/(19000-50)*100\% \quad (2)$$

where Cl^-_{sam} is the Cl^- concentration of borehole water. The mixing ratio between the sea and recycled water was calculated using equation (3).

$$M_{sea2} = (Cl^-_{sam} -450)/(19000-450)*100\% \quad (3)$$

The actual value is between M_{sea1} and M_{sea2}. The difference between M_{sea1} and M_{sea2} was very small and the average value was adopted. Nitrate (NO_3^-) is also an important indicator because the sea and river concentrations are very low relative to the recycled water, and NO_3^- concentration is very sensitive to the mixing ratio of the recycled water.

The mixing ratio between the recycled water and river water (recharge water) was calculated using equation (4).

$$M_{recy} = NO_3{}^-_{sam} /100*100\% \quad (4)$$

where $NO_3{}^-_{sam}$ is NO_3^- concentration of borehole water. If M_{recy} was less than 100 %, then M_{recy} value was adopted as a mixing ratio of the recycled water. However, when the M_{recy} value was over 100 %, this value cannot be used. In that case, the groundwater is thought to be contaminated by nitrogen fertilizer from farm land on the delta. When both the M_{sea} and M_{recy} values can be derived, all mixing ratio of sea, recycled, and river waters can be determined. The accuracy of this process was confirmed by isotope values. The difference between the calculated groundwater oxygen isotopic ratio from the determined mixing ratio of sea (0 ‰), recycled water (-3.3 ‰), and river water (-5.8 ‰), and the measured isotopic ratio was compared. When the difference between both the isotopic ratios was less than 2 ‰, the mixing ratio was fixed. When the M_{recy} is over 100 % or the difference is over 2 ‰, mixing ratio cannot be determined.

RESULTS AND DISCUSSION

Climate and Rainfall Pattern

Analysis of climate data (from BOM stations, Figure 1) showed rainfall distribution is characterized by longer periods of wetness, mostly between June and October, with less rainfall in late summer and autumn of each year (January-May) (Fig. 3).

This seasonal drought pattern is sometimes punctuated by occasional, high intensity storm events in summer, e.g. the amount of rainfall in January 2011 was between 111 mm and 211 mm across all the stations. Average annual rainfall within the catchment is highly variable, decreasing from annual (1950-2011) mean of 990 mm in the upper catchment (around Blackwood) to 483 mm (in Melton) and towards the south-west coast around Werribee and Little River (530 mm).

The rainfall distribution is considered unimodal in most of the stations in the catchment with peaks around July/August (Figure 3a-h), although Laverton/Werribee showed a relatively even distribution throughout the year. The long-term rainfall data recorded at Melton, Little River, Ballan, Blackwood and nearby Melbourne BOM Stations (with station numbers indicated in Figure 1) since the early 1950s to present indicated that the highest rainfall also occurs in winter and spring.

These long term sequences of rainfall at stations within the Werribee River basin were examined to understand the natural variation of rainfall including the cycle of wet and dry periods which may better describe rainfall processes.

The seasonal patterns of rainfall and temperature in the basin are shown in Figure 3. Monthly average temperatures range from 25-32°C in summer to 3-14°C in winter. Analysis of the rainfall data reveal the Werribee, Little River, Melton, Ballan and Blackwood areas were subject to a cycle of higher and lower rainfall periods.

The impacts of drought and perhaps climate change on the area are demonstrated by an increase in temperature by 0.4 C during the decade 1998 to 2007 in comparison to a thirty (1961 to 1990) year average, and a 14% decrease in annual rainfall (DSE 2008).

Hydrological Data

The water authority (Southern Rural Water, SRW) monitors the streamflows into the reservoir, the volume of water that is diverted into the reservoir from nearby creeks, extractions to meet urban demands, and the compliance with environmental flow rules. Detailed hydrological data with charts for all Werribee Catchment rivers and streams (e.g. Werribee River, Lerderderg River, Parwan Creek, Toolern Creek) are available online (http://www.melbournewater.com.au/content/rivers_and_creeks/rainfall_and_river_level_data/rainfall_and_river_level_data.asp) and updated regularly. The Werribee River is characterized by low flow for most of each year, except during the winter and spring seasons when heavy rainfall causes a relatively higher discharge. The river hydrograph is highly controlled by the intensity and duration of precipitation during the winter season (Figure 4). Reduced flows have been exacerbated by the recent prolonged drought (1997-2007) as documented in LeBlanc et al (2009). Ephemeral streams or creeks, which mostly discharge into the Werribee, Lerderderg and Little rivers, begin to flow towards the end of May each year and ceases to flow by early December. Toolern Creek and the lower Werribee River have increased levels of low flow due to private diversions and also showed summer and winter stresses (Figure 4). About 7.4×10^6 m^3 water of the Lerderderg River is annually discharged into the Werribee River, of which ~80 % of annual discharge flows during the winter and spring seasons (DSE 2008). There are three major dams, notably Pykes Creek, Merrimu and Melton Reservoirs. Flow regimes are highly regulated and water is extracted for irrigation and domestic use (SRW 2009).

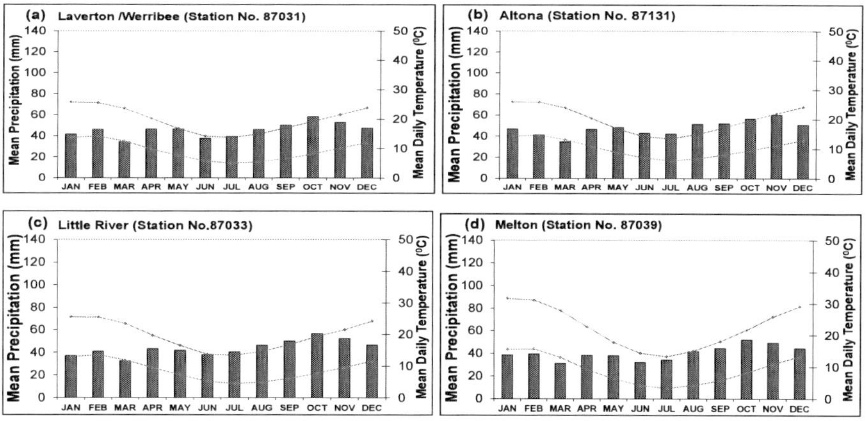

Figure 3. (Continued).

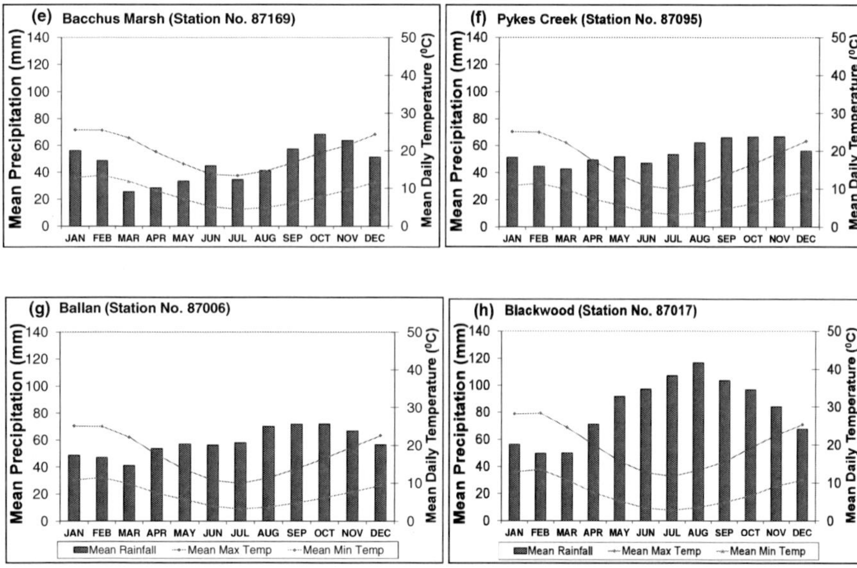

Figure 3. (a-h): Mean monthly maximum (red line with square symbol) and minimum (purple line with triangle symbol) air temperature (° C) with mean monthly rainfall (mm, shown as blue bar) at meteorological stations of the Werribee River Basin. All station locations are shown in figure 1 (Data source: BOM online, www.bom.au).

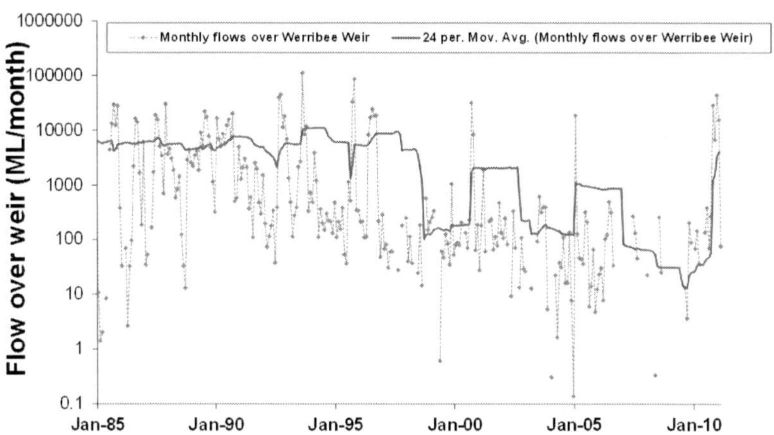

Figure 4. Werribee River monthly flow characteristics (data from the SRW).

Stable Isotope Hydrology

Isotopic Composition of Precipitation

The pattern of rainfall in Melbourne is similar to that of Werribee and other locations within the basin (except at Blackwood, Figure 2). Since there was limited isotopic analysis undertaken for rainfall at Werribee site, the $\delta^2H/\delta^{18}O$ data monitored at Melbourne and surrounding suburbs including Melton (IAEA 2006) was used to produce the LMWL for the study area (Figure 5). This isotopic composition of precipitation is in the range -13.0 to -0.1‰ $\delta^{18}O$ and -94.7 to 4.1‰ δD.

The weighted average values of $\delta^{18}O$ and δD in precipitation for this area are -4.9 and -26.0‰, respectively. The deuterium values are linearly and positively correlated to the $\delta^{18}O$ values (Figure 5a), with the equation for the best-fit line as follows:

$$\delta D = 6.98\delta^{18}O + 6.8 \ (r^2=0.92) \qquad (5)$$

The basis for this kind of correlation, generally, lies in the fractionation of isotopes during evaporation-condensation processes (Gourcy et al., 2005). Comparing the LMWL developed by Crosbie et al (2012) for Melbourne ($\delta D = 7.5\ \delta^{18}O + 9.8$), the slope of the LMWL developed for the Werribee basin is lower as it is located inland (20-40 km) northwest of Melbourne where the climate is relatively drier (except at the far north-end of the basin around Blackwood, approximately 60 km from Melbourne; Figure 3h). The LMWL also differs from the Global Meteoric Water line (GMWL, $\delta D = 8\ \delta^{18}O + 10$, Rozanski, et al., 1993). It has an isotopic composition with lower slopes, lower intercepts and the data show higher values for deuterium excess when compared to the GMWL.

Evaporation during rainfall would generally shift the isotopic values away from the GMWL (Friedman et al., 1962) and this effect is greatest for light rains (Clark and Fritz 1997). Figure 5b shows that the slope of the LMWL for rainfall (>40 mm/month) became is closer to that of the GMWL and agrees with results of Yin et al (2011). Earlier, Clark (1987, as reported in Mook (2006)) postulated that rainfall events greater than 20 mm oriented the slope closer to 8 (i.e. slope of GMWL); but Yin et al. (2011) regarded this as an underestimation and demonstrated that the threshold of precipitation for secondary evaporation to have no effect is >40 mm/month.

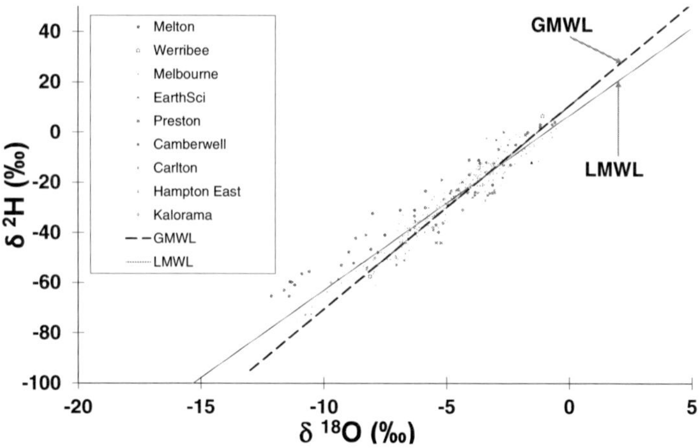

(a)

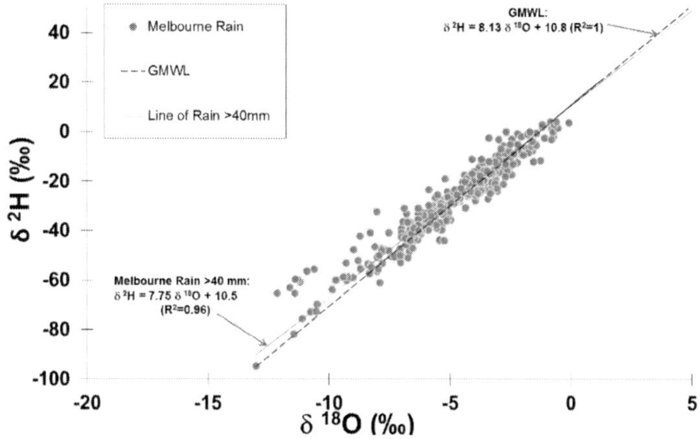

(b)

Figure 5. *a:* The isotopic compositions of precipitation samples and their resultant LMWL. *b*: Isotopic composition of precipitation events greater than 40 mm/month (Note: the blue dots as local meteoric water measurements).

The amount-weighted average δ^2H and δ^{18}O composition of rainfall reported by Crosbie et al (2012) for a two-year period (2008-2010) at Melbourne are as follows: δ^{18}O (-5.60) *δD* (-31.84) d-excess (12.96) slope of LMWL (7.5) intercept of LMWL (9.8). This rainfall isotopes sampling at Melbourne coincides with our study period (with selected seasonal climate and isotope data in Table 3) and the values (i.e. slopes, intercepts and deuterium excess) of precipitation suggest addition of re-evaporated moisture from a continental source mixing with water vapour that has travelled inland from the coast representing the isotopic composition for the basin. Similar data and conclusions have been drawn for other parts of Australia (Crosbie et al., 2012; Hughes & Crawford 2012) and outside Australia (Ingraham 1998; Gat et al., 2001, Mazor 2004, Gourcy 2005, Mook 2006) . suggesting re-evaporated moisture with waater vapor that has travelled inland from the coast.

Clark and Fritz (1997) stated that secondary evaporation also decreases the deuterium excess (*d*) defined as:

$$d = \delta D\text{-}8\delta^{18}\mathrm{O} \qquad (6)$$

Our study show *d* values in rainfall are >10 with an average of 10.9 (1962-1998) and 12.4 (2008-2010).

Table 3. The δ^{18}O and δD compositions of rainfall sampled at Werribee during the study period (2007-2010)

Sample Period	Rainfall (mm)	Temperature (°C)	Humidity (%)	δ^{18}O (‰)	δD (‰)
December 2007	69.8	41(26)	48.0	-3.70	-14.15
September 2009	58.8	31(19)	53.0	-6.85	-43.50
October 2009	33.8	30(19)	54.0	-8.13	-57.56
February 2010	27.2	35(27)	50.0	-3.60	-20.7
April 2010	19.2	31(23)	55.0	-1.10	6.70

Note: Temperature indicated is monthly maximum while the mean of maximum temperatures for the month is in parentheses.

Stable Isotopes Composition of River Water

In river water, δ^{18}O values range between -5.85 and 0.3‰, and δD between -36 and 7.1‰, respectively. There are differences in isotopic composition between the Werribee River and other catchment rivers and creeks contributing to the river (i.e. Werribee River tributaries). For instance, the average values of δ^{18}O and δD are -0.98 and -11.55‰, respectively, for the Werribee River while the average values for samples from its tributaries wereas -3.2‰ for δ^{18}O and -18.1‰ for δD during the summer of 2010 (Table 4). Winter samples have similar average values (Werribee River: -2.3 δ^{18}O and -17.5‰ δD; Werribee River tributaries: -3.2 δ^{18}O and -17.5‰ δD). While differences are equally observed between the summer and winter seasons for Werribee River samples, there is no seasonal variation in isotopic values for the tributaries (i.e. selected rivers and creeks within the basin).

Using the least square method, the relationship between δD and $\delta^{18}O$ was calculated as δD=3.9δ ^{18}O - 11.3 (r^2 = 0.98) for the Werribee River (30 samples) and δD=4.65δ ^{18}O - 2.51 (r^2 = 0.85) for the tributaries (Figure 6). Compared with the relationship between δD and $\delta^{18}O$ in samples from the tributaries of the Werribee River, most of the samples from the Werribee River are scattered below the LMWL, defining an evaporation lines (Figure 6). Studies have shown that a significant percentage of water can be evaporated from in-stream river water, particularly in rivers with reservoir storages, as demonstrated for the Murray River (Australia), and other rivers in arid to semi-arid regions (Simpson et al., 1991, Vrbka et al., 1993, Yao et al., 2009). Some of the Werribee River tributaries samples indicated that river water has undergone a greater degree of enrichment.

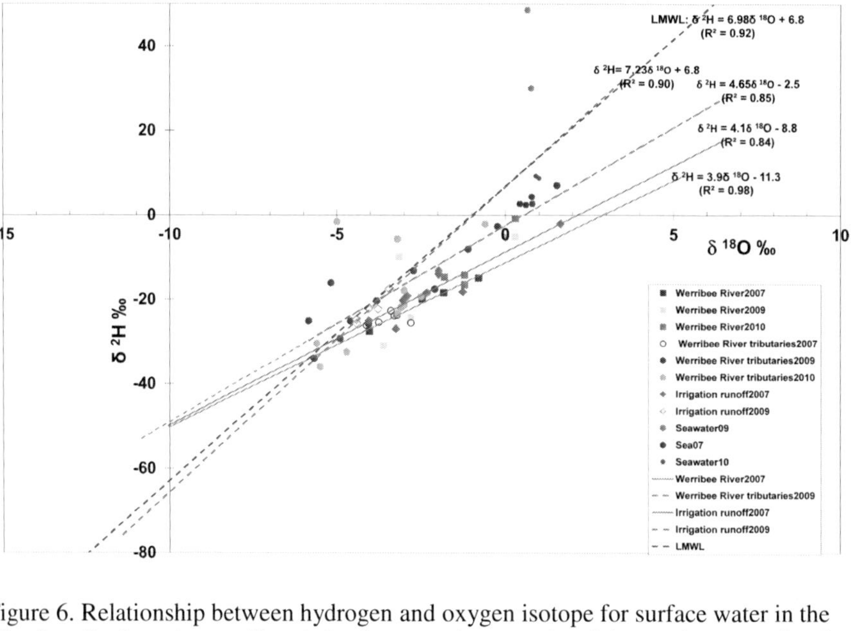

Figure 6. Relationship between hydrogen and oxygen isotope for surface water in the Werribee Basin and water from irrigation near the central and the southern margin of the basin.

Irrigation runoff water showed similar patterns to river water (Figure 6) deviating from the LMWL with a slope of 3.9 relative to 4.16 for river water,, indicating that it has undergone a greater degree of evaporation thanirrigation water is river-water. However, the 2009 samples of irrigation runoff were very close to the LMWL showing enrichment in lighter isotopes influenced by the winter rainfall (and a wider range from summer samples).

Sea water samples plot to the right of VSMOW ($\delta^{18}O$ and δD of 0‰ and 0‰, respectively) as seen in Figure 6, indicating some minor fractionation by evaporation in the estuary.

The $\delta^{18}O$ values of water from the Werribee River and its tributaries show a range from -5.85 to +1.53 ‰ during both winter and summer seasons, indicating that rivers of the basin derive their water from common origin. This is a wide range for $\delta^{18}O$ and indicates a mixture of meteoric derived surface runoff, subject to variable degrees of evaporative concentration, and with possibly some groundwater contribution.

The overlap indicatesing connections of river water, groundwater and irrigation water and possibly groundwater, although no strong evidence in support of groundwater (Figure 7). Except near the river's source (northeast of Ballan) and the Lerderderg River at O'Briens (upper catchment tributary), where δD values appear to be more depleted,; $\delta^{18}O$ values for the Werribee River vary in a narrow range of -4.70 to -0.00 ‰, indicating that the river has a similar water source. In the lower segment of the basin, the river water indicates mixtures of various proportions of different sources. Ii et al. (2012) proposed mixing of river water, groundwater, recycled water and seawater in the study area in various proportions using conservative indicators e.g. chloride and fluoride. In summer, large volumes of water are applied for irrigation in the basin, especially in the WID and BMID. By implication, the residence time of water is not stored in the reservoirs during this period is reduced but as it is mostly channelled into the irrigation districts, with the result that the isotopic composition of river water in both upper and lower catchments is similar. During winter, river water is stored in the reservoirs, in which case, the stored water would be enriched in D and ^{18}O relative to the upper catchment because of evaporation, causing river water in basin to plot below the LMWL. Water from Lerderderg River, located to the northwest of Bacchus Marsh, sampled in summer of 2010 (February) plotted farther above the LGMWL, just suggesting a significant influence of inland moisture recycling of evaporated water as most precipitation samples (Fig. 5a). This and the Werribee River sampled at Ballan plot above the LMWL.This and the Werribee River sampled at Ballan plot above the LMWL.but all the summer 2007 samples plot on and below the LMWL. All other samples collected in summer of 2007 and 2010 plot on and below the meteoric line.

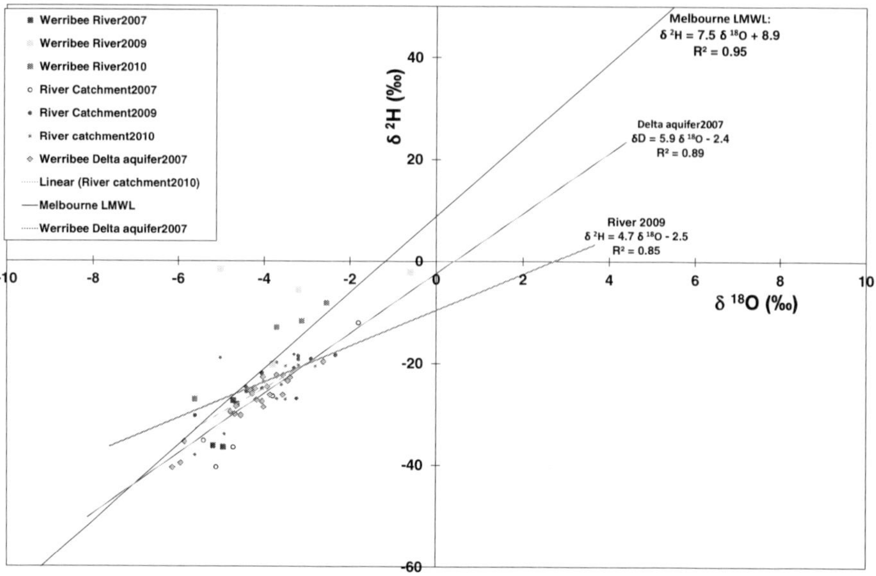

Figure 7. Plots of all surface water in relation to precipitation, irrigation water and groundwater samples during the study period. (Note: River Catchment in the legend refers Werribee River tributaries. Also, the year of sampling are indicated in the legend to represent different seasons – 2007 and 2010 represents summer sampling (Nov-Jan) while 2009 represents winter sampling (Jun-Aug)).

Stable Isotopes Composition of Groundwater

The distribution of $\delta^{18}O$ of both shallow and deep groundwater ranges from -6.70 to -1.29 ‰, and δD from -43.86 to 1.49 ‰. The mean values of $\delta^{18}O$ and δD are -4.20 and -25.10 ‰, respectively. A plot of δD against $\delta^{18}O$ of all analyzed groundwater samples shows linearly and positively correlated values (Figure 8), with linear regression represented by the following equation:

$$\delta D = 6.9\ \delta^{18}O + 4.3\ (r^2 = 0.63) \quad (7)$$

The plot of $\delta^{18}O$ and δD values also shows two distinct groups (A and B), with group A representing all 2009 sampling (July-August, winter season) while group comprise mostly of 2007 and 2010 summer (Dec/Jan) samples (Figure 8).

Group A are relatively highersamples lie mostly above the LMWL and have a broader distribution. Generally, group B is isotopically depleted relative to group A, although a greater number of basalt aquifer samples appear to be more depleted in group B compared to A. Group A mostly lie above the LMWL (and GMWL) while group B lies below the line with a number of theits samplesm defining the evaporation line described in figure 9.

In general, the data show that groundwater (samples collected in summer of 2007 and 2010) plots below the LMWL. The δD and δ^{18}O values of shallow groundwater is similar to values obtained for the Werribee River tributaries (mean δD and δ^{18}O values are -24.4 and -3.4‰) (Figure 7). In the basalts, six groundwater samples in 2007 and three samples from the Delta aquifers are much more depleted in D and ^{18}O than rivers of the catchment, which indicates that groundwater receives contributions from a source relatively depleted in D and ^{18}O, instead of from upper catchment rivers. Four bore samples, collected in the upper catchment (around the steep gorge near Bacchus Marsh) where the elevation is relatively higher, are depleted in D and ^{18}O, relative to the rivers (Werribee, Lederderg, and Parwaon creek). This probably relates to more rapid recharge by heavy rain events in these areas and a relative lack of evaporation (assumes recharge by river).

Although tThe majority of shallow groundwater samples were taken in bores less than 30 m in depth (Table 2). Samples taken in bores from 31-70 m were classified as intermediate (i.e. medium deep) groundwater while samples taken in bores deeper than 70 m were regarded as deep groundwater. There are seven bores in the category of the deep groundwater; these are mostly private bores.

Saline groundwater is found locally within 10 km from the coast; this groundwater can has been described as a mixture of fresh recharge, recycled water and a components (maximum 60%) of marine water (Ii et al., 2012). While some groundwater over 9 km from coast showed high sea water mixing ratio (20%), the sea water mixing ratio calculated for groundwater beyond 10 km from the coast (i.e. borehole medium and borehole upstream, Table 2a) was <10% (Ii et al., 2012).

Similar observations are reported in coastal groundwater nearby (Western Port Basin) south-east of Melbourne where the marine component of groundwater in samples near the coast is up to 20%, shown by Cl-δ^{18}O mixing relationships (Currell et al., 2013).

Groundwater total dissolved solids (TDS) concentrations range between 290 mg/L and 24,681 mg/L (Allinson et al., 2011, unpublished data). The freshest groundwater is from the Brighton Group aquifer around Bacchus Marsh (bores G-63, 65, 68, and 69), which ranges from 499-852 mg/L TDS (median 693 mg/L) through the sampling period. This is followed by seven bores screened in the delta aquifer around the central part of the WID, especially from southern end to northern end of K road (TDS 894-1053 mg/L, median 998 mg/L; in with the exception of bore G-70 that showed 290 and 367 mg/L, respectively for summer 2010 and winter 2009). And then followed by gGroundwater from the basalt aquifer has moderate-high salinity (TDS 922 to 10,280 mg/L, mean 2,820 mg/L).

While shallow groundwater (<30 m depth) is generally the most saline, the highest values are from bore G-21 (16,585-24,681 mg/L, median 22,469 mg/L for the four years of sampling). This bore is located in the central eastern coast and screened between 26-28 m bgs in the delta aquifer. It is contrasted with two other bores in the same location (G-20 and G-37) but screened at different depths (7-12.8 and 13-22.5 m, respectively). These two have similar TDS (2804 and 2814 mg/L, respectively). This indicates the lateral encroachment of sea water at a depth of approximately 30 m and rules out vertical mixing from water contaminated by modern agriculture (Ii et al., 2012). Therefore, the source of high salinity in is the sea; historically the bore site G-21 has been subjected to sub-tidal events and water level has declined to below sea level since 1997 (SRW 2009, unpublished data). The median of TDS values for the deep bores (screened between 169-260 m bgs) in summer and winter are 2212 and 2170 mg/L, respectively (range is between 2160 and 2260 mg/L). These are bores located along the same flow path in the NW end of Deutgam WSPA, except G906 located outside the WSPA (around Sanctuary Lakes Golf course).

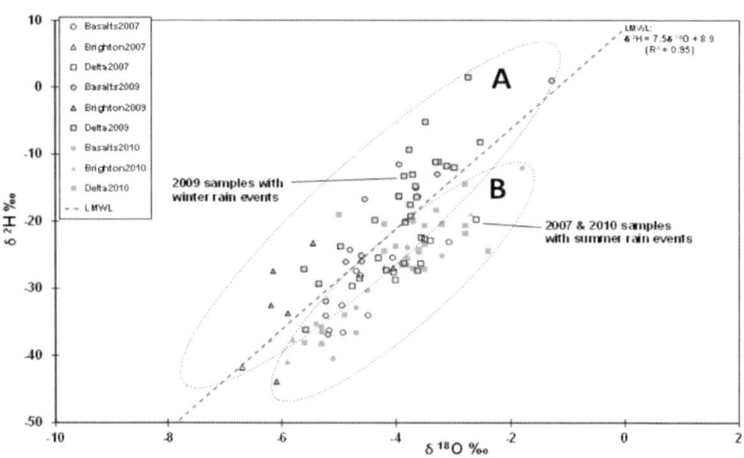

Figure 8. Relationship between hydrogen and oxygen isotopes for groundwater in the study area (*Group A* represent groundwater sampled in winter 2009 and *Group B* represent samples of summer 2007 and 2010).

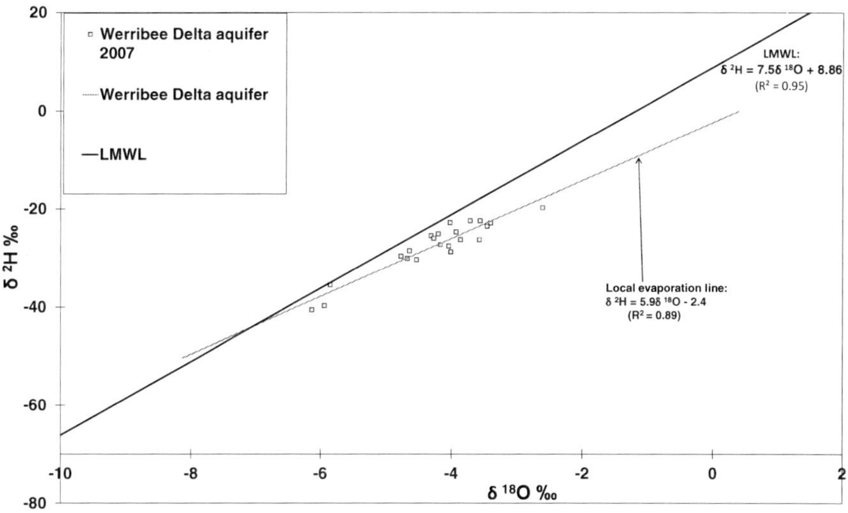

Figure 9. Local evaporation line of the Werribee delta aquifer groundwater.

The isotope composition of these bores are as follows: $\delta^{18}O$ range from -6.2 and -5.9, δD range from -40.9 to -37.7 (in summer); while the winter sampling showed values of -43.9 and -27.4 for δD and -6.2 to -6.1 for $\delta^{18}O$. These isotope values are quite depleted, and is common for deep, old groundwater samples to have depleted stable isotopic composition, due to recharge in the cooler late Pleistocene period.

This has been demonstrated data from the Westernport (south-east of Melbourne) (Currell et al., 2010, 2013).

Vertical Mixing and Origin of Groundwater

Evidence of vertical mixing of groundwater is shown by looking at the vertical distribution of nitrate concentrations. (Figure 10). The major N source in the catchment is modern agriculture (Wei et al., 2013); due to the lack of natural nitrate in most geologic formations, NO_3 concentrations >5 mg/L generally indicate contamination by fertilizers, animal wastes and/or effluents (Heaton, 1986). High NO_3 concentrations in deep groundwater are therefore a strong indication of downwards vertical leakage of shallow groundwater resulting from anthropogenic contamination from the surface (e.g. Currell et al., 2010). The deep groundwater samples (17 samples from 7 bores) generally have low nitrate concentrations (median 0.1 mg/L); while the majority of shallow groundwater samples (0.1-30 m depth) have elevated NO_3 concentrations between 0.1 and 336 mg/L. Of the intermediate depth samples (31-70 m), 10 of 50 have >45 mg/L nitrate, all in the basalts, indicating that vertical leakage has occurred to depths beyond 30 m. All but two of these samples are within the WID in the south-west corner of the study basin (Figure 1). More samples are needed to confirm nitrate contamination in the BMID, although there is evidence that the Rowsley Fault (around Bacchus Marsh) could be acting as a preferential pathway for downward leakage. Nitrate concentration in bore Gs-69 (screened in the Brighton Group) was up to 32 mgNO_3/L during the January 2010 sampling (and 18.3 mg/L during winter, August 2009) as compared to Gs-63 with <1 mgNO_3/L throughout the study period. Vertical leakage is also consistent with higher dissolved oxygen values in these samples compared to the rest of the seven deep groundwater in Brighton Group (2.6 and 2.4, respectively in Gs-69 compared to 0.1 mg/L in Gs-63). There is also evidence of more gradual vertical mixing occurring in the basin from the TDS distribution; for example, a number of samples showing elevated NO_3 concentrations from intermediate boreholes (31-70 m) have elevated TDS contents, and TDS and NO_3 contents generally decrease and homogenizes with depth (Figure 10a, b). It is important to note that the elevated salinity in the basalt at 30 m is a special casewas attributed due to lateral seawater intrusion. These data are consistent with gradual vertical mixing between saline shallow borehole delta and lower salinity deeper groundwater in the Brighton Group aquifer and the calculated mixing ratios (Ii et al., 2012). The surface water evaporation line(s) plotted on the same graph with groundwater shows some overlap. The overlap may be caused by (1) evaporated surface water used as an irrigation source recharging the aquifer locally; (2) evaporated groundwater providing baseflow to the rivers (although from isotope data this is less likely). Even though we cannot, at this stage, distinguish which is more important, evidence from field observation during sampling showed that during periods of high temperature and strong northerly winds, irrigation is used almost continuously to prevent wilting and stressing of leaf crops such as lettuce. The extensive recharge this generates over a few days in summer will be highly subjected to evaporation; hence surface water leakage is likely to influence shallow groundwater within the irrigation district. Overall the data presented shows extensive recharge by rainfall, but the surface water leakage could be an additional source as evident in Ii et al (2012) mixing study. Isotopic data showed shallow groundwater is also be composed of precipitation from the winter and springs seasons, as evidenced by the variability in groundwater isotopes observed in different sample campaigns (Figure 8). Some shallow groundwater must have been recharged following interaction with soils containing high N contents from fertilizers and has likely also undergone transpiration via crops before reaching the saturated zone – as evidenced by elevated salinities but still meteoric stable isotopic contents (transpiration is a non-fractionating process). The δD and $\delta^{18}O$ values in shallow saline water (Figure 12) being similar to recent precipitation are consistent with this mechanism. The fact that shallow groundwater has a significant modern precipitation component and that irrigated agriculture has been practiced in the region for several decades, means that a large proportion of shallow groundwater would have been recharged since intensive cropping of shallow rooted crops and fertilization began. The high groundwater TDS and NO_3 concentrations in this water (Table 4; Figure 10) may also indicate a component of irrigation returns (e.g. surface water, recycled water) that have undergone transpiration and dissolution of N from fertilizer applications.

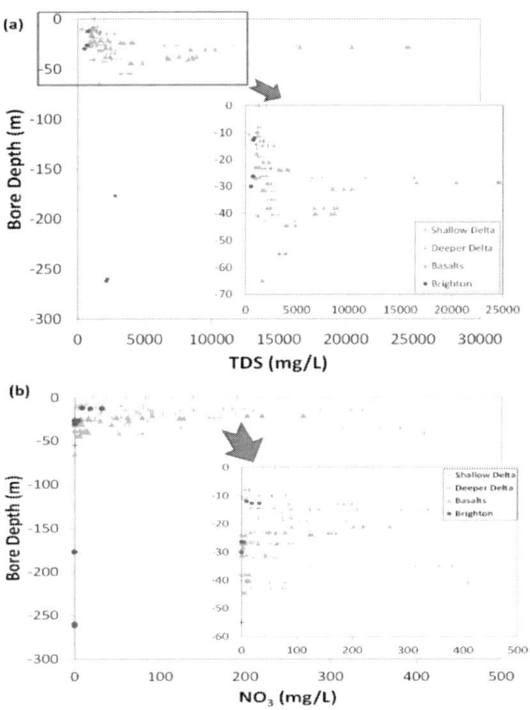

Figure 10. Plots showing the variation with depth of: (a) TDS content of groundwater and (b) groundwater nitrate concentration.

Irrigation returns might be expected to have low $\delta\,^{2}D$ and $\delta\,^{18}O$ values as fresh groundwater, although since 2004 groundwater has been mixed with surface and recycled water in order to reduce the high salinity of the water used for irrigation. This has affected the composition of irrigation water (and the overall contribution of each component fluctuates annually). The $\delta\,^{18}O$ of the recycled water was similar to that of borehole delta waters. The recycled water mixing ratio estimated in Ii et al (2012) was uniform with depth (10 to 30 % at all depths), suggesting that recycled water had infiltrated uniformly down through the permeable Werribee Delta soils.

The stable isotope data of groundwater for most of the basin, especially the south-west coast, together with major ion compositions (Allinson et al., 2011, unpublished data) suggest two marine sources of solutes: (1). marine aerosols delivered in rainfall; (2) direct seawater intrusion for intermediate or deeper groundwater in addition to solutes derived from fractured basalts, dissolution or attributed to mixing. Both the $\delta^{18}O/Cl^{-}$ and Cl^{-}/Br^{-} ratios imply that evaporation (e.g. from shallow watertable, Table 2a) and transpiration (e.g. from all-year-round cropping) during recharge represent the processes controlling groundwater TDS contents locally. Shallow groundwater, particularly from the delta aquifer, must have been derived from a combination of sources: the sea, recycled water and river delta water from surface irrigation and upstream groundwater from the upper catchment area.

Salinization and Evaporative Loss

Salinity (as measured by the EC or TDS of groundwater) from historic data (SRW, 2009) to recent measurements reveals the varying quality of groundwater in the Werribee River catchment as illustrated in the time series plots (Figure 11a-f). These figures show the variation of groundwater salinity and Werribee delta aquifer drawdown; during the early 1990s seasonal drawdown was <0.5 m with recovery during summer. However, since 1996 recovery has shifted to winter and the seasonal drop in watertable increased up to 2 m, which indicates heavy reliance on groundwater due to the inadequate supply of irrigation water from Werribee River during the extended period of drought.

The spatial distribution of salinity across the WID shows that the northern and eastern parts of the districts have relatively higher salinity than the central-southern part (SRW, 2009),; and from recent data groundwater is generally more saline in the area surrounding the WID (south-west coastal areas of the basin) compared with regions further to the north-west and east. The highest groundwater salinity was recorded in bore G-73, located closest to the coastline. The source of this salinity in the Werribee Delta has been traced to the influence from seawater (Adelana, 2011; Allinson et al., 2011).

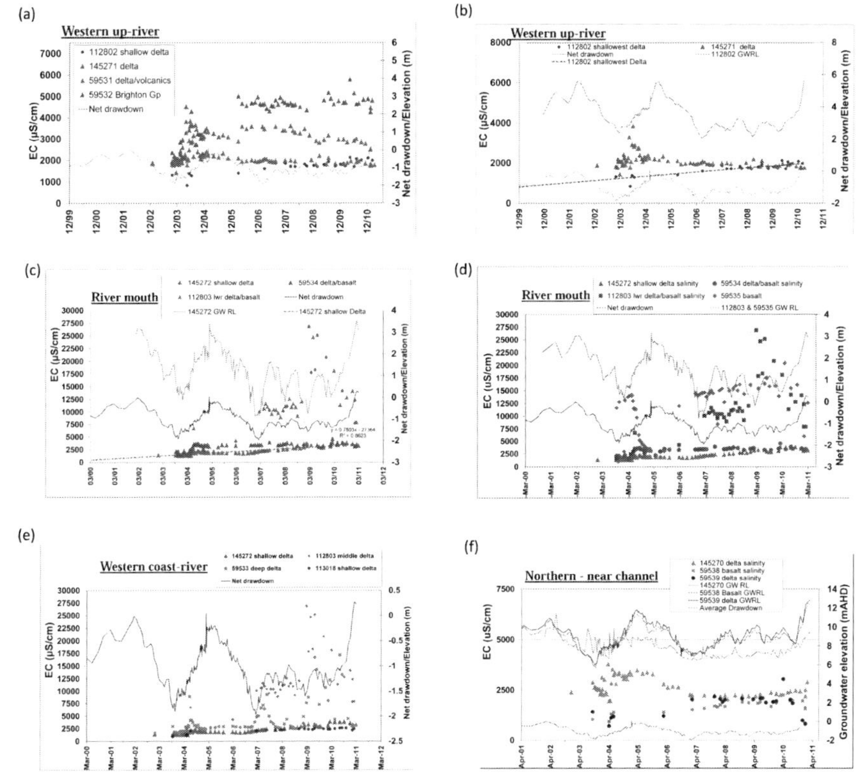

Figure 11. Spatial and temporal variations of groundwater salinity in the Werribee Basin. Note: Net drawdown/seasonal water table fluctuations (Note: the numbers e.g. 145272 represent SOBN bore; RL is reduced level).

In addition to the salinity from resulting from transpiration by crops and other sources discussed, a plot of chloride versus $\delta^{18}O$ (Figure 12) help to explain the mechanism of salinization in the study area. Two water types can be identified; the first type is waters in which dissolution of soluble salts leads to salinization. In this water type, an increase in chloride concentration does not correspond with isotopic enrichment of $\delta^{18}O$. Most of the bores tapping the three aquifers (Werribee delta, basalt and Brighton Group) belong to this group (except five samples from the January 2010 sampling, which are affected by evaporation). The other water type shows an increase in chloride concentration in this water type corresponds with a gradual increase in $\delta^{18}O$

composition. This implies that dissolution of soluble salts or mineral weathering could be one of the contributory factors for groundwater salinity, whereas evaporation could also account for salinity based on shallow bores with elevated water table in the study area.

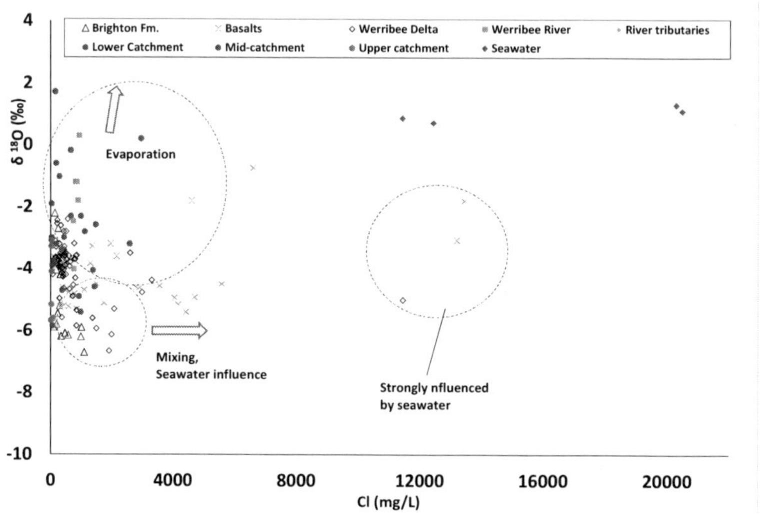

Figure 12. Plot showing the relationship between chloride (Cl) concentration and isotopic composition of oxygen ($\delta^{18}O$) for groundwater samples in the study area. (Note: The samples in the red dotted circle is showing strong seawater influence)

Table 4. Isotopic composition of groundwater in delta and basalt aquifers (representing the shallower groundwater)

Sample ID	Easting GDA 94z55	Northing GDA 94z55	Date	Cond. (µS/cm)	Temp	18O (‰)	2H (‰)	d (‰)
G06-1	297593	5794644	26/11/2007	2940	16.9	-3.9	-24.7	7.6
G37-1	295833	5795434	26/11/2007	9760	16.6	-4.8	-29.6	9.5
G38-1	297483	5794976	26/11/2007	1635	16.9	-3.6	-22.4	6.8
G39-1	297483	5794974	26/11/2007	2197	18.3	-3.4	-22.9	5.0
G02-1	295413	5798184	27/11/2007	1994	18	-4.0	-27.5	5.6
G31-1	297673	5796784	27/11/2007	1542	17.7	-2.6	-19.7	1.7
G72-1	295413	5798184	27/11/2007	3400	18.2	-5.9	-35.5	12.5
G73-1	295413	5798184	27/11/2007	4700	18.2	-5.9	-39.7	9.0
G04-1	300293	5798164	28/11/2007	2235	17.3	-3.5	-23.5	4.8
G33-1	300293	5798164	28/11/2007	4210	17.3	-3.6	-26.3	3.0
G34-1	297853	5798434	28/11/2007	1715	16.7	-4.0	-28.7	4.2

Sample ID	Easting GDA 94z55	Northing GDA 94z55	Date	Cond. (μS/cm)	Temp	18O (‰)	2H (‰)	d (‰)
G18-1	300293	5798164	28/11/2007	6590	19.1	-6.1	-40.6	9.7
G35-1	297193	5799764	29/11/2007	2168	16.7	-3.9	-26.2	5.4
G36-1	295613	5797984	29/11/2007	1855	17.8	-4.2	-27.2	7.0
G32-1	299213	5796084	30/11/2007	4390	16.5	-4.3	-25.4	9.9
G71-1	299213	5796084	30/11/2007	3690	16.8	-4.5	-30.3	6.8
G03-1	295593	5795564	4/12/2007	2374	17.2	-4.6	-28.5	9.5
G86-1	295593	5795564	4/12/2007	3430	16.8	-4.7	-30.1	8.2
Pb1-1	296760	5798206	6/12/2007	1978	16.3	-4.0	-22.8	10.2
Pb2-1	296173	5798761	6/12/2007	2455	17.2	-3.7	-22.4	8.0
Pb3-1	296108	5799720	6/12/2007	2960	17	-4.2	-25.0	9.4
Pb4-1	297548	5795202	6/12/2007	2670	16.5	-4.3	-26.0	9.0
Pb5-1	294690	5801040	6/12/2007	4380	28.4	-6.7	-41.8	13.2

Overall the sea water mixing ratio was high in the deeper groundwater close to the coast (e.g. at 30 m in depth; illustrated in Ii et al., 2012), and this zone is thought to be a salt-water wedge (Ii et al., 2012). The mixing ratio of recycled water varied from 0 to 60%, but was mainly in the range 10 to 30%. This mixing ratio (%) range of recycled water was good in agreement with isotopic data.

Evaporation (either during recharge or from shallow water tables) or influx of saline water from the sea and estuary should cause systematic displacement from the meteoric water lines (Clark & Fritz, 1997; Cartwright et al., 2009), increasing $\delta^{18}O$ values as Cl concentrations increase (Figure 12). The twenty-three samples from delta aquifer (Table 4) used to develop the local evaporation line (Figure 9a) defined by $\delta D = 5.9\delta\ ^{18}O - 2.39$ (r^2=0. 89) is an example of such displacement. The intercept of the evaporation line and the LMWL is the original isotopic composition of groundwater before evaporation i.e. -5.5 and -35.2‰ for $\delta^{18}O$ and δD respectively and is quite close to the average values of precipitation ($\delta^{18}O = -5.3‰$ and $\delta D = -29.7‰$) for the period 2008-2010 (Crosbie et al., 2012). The slope of the line is 5.9, indicating groundwater had experienced some evaporative loss.

Recharge Mechanism and Groundwater Flow

In the Werribee basin, the $\delta^{18}O$ – δD composition of groundwater in summer plots below but almost parallel to the LMWL. This deviation can be used to explain the recharge mechanism in the basin, i.e. that soil water within the unsaturated zone evaporates leaving a slightly isotopically depleted soil moisture.

The evaporated soil moisture mixes with subsequent rain, which infiltrates into the soil and displaces the residual soil water downward. Eventually, this mixed water will reach the water table as recharge (Allison et al. 1984). In this way the isotopic composition of such recharge waters will then be a mixture of pre-existing soil moisture and modern rainfall.

As this process continues repeatedly the result is a groundwater with a range of isotopic compositions controlled by isotopic variation in the rains and the loss of soil moisture by evaporation (Clark & Fritz, 1997). If the recharge conditions remain relatively uniform over time, groundwater should follow a line parallel but displaced from the LMWL as demonstrated in figure 9 and discussed under groundwater isotopic composition above. Detailed explanation for this enhanced evaporation effect leading to isotopic variations and recharge characteristics are given in Allison (1984) and (Clark & Fritz (1997).

Further evidence for recharge processes in the study area can be illustrated in a plot of *d* versus $\delta^{18}O$ (Figure 13). It is obvious from this figure that the Brighton water, depicting deep groundwater, has lower delta O-18 than the basalt and Werribee delta water but the differences in d-excess are the same. The deuterium-excess shift towards that of the rainfall (in Figure 13) suggests a modern recharge. This is further evident in from the modern rainfall δD and $\delta^{18}O$ values presented in section 4.3.1.

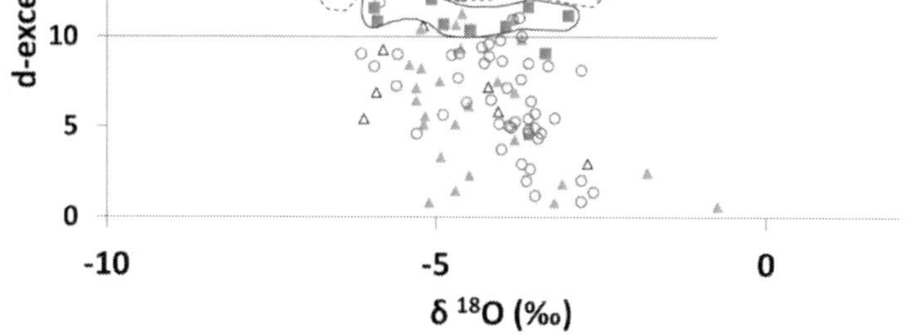

Figure 13. Comparison of deuterium excess with $\delta^{18}O$ in rainwater (2007-2010) and bores screened in the various (Delta, Basalt and Brighton Formation) aquifers.

However, the fact that Brighton Group aquifer samples are mostly depleted isotopically relative to the other aquifers in either group B or group A (Figure 8) suggest that deep groundwater may have been recharged during cool periods. This suggests that the Brighton water was recharged under different conditions – lower end of the d-O-18. This could have been due to cooler climate (i.e older water) and indicates that recharge to the Brighton Group is occurring, much more slowly than to the surface aquifer, i.e. connectivity to the Brighton Group is relatively restricted. or due to different recharge zone – e.g. closer to the divide. Radioisotope dating of the waters in this region would help to confirm the groundwater age.

MANAGEMENT IMPLICATIONS

Groundwater from the Werribee Delta aquifer is a viable resource and using isotope geochemistry to identify where potential surface contamination may occur or confirm seawater intrusion is an important contribution to protecting valuable groundwater resources within the Local Government areas near the metropolitan city of Melbourne. Under the Southern Rural Water plan on the WID, the key driver of groundwater management is to avoid drawing down the aquifer to the point where seawater intrusion takes place. Several studies in line with the management strategy have therefore been in place since for the last 10 years (SRWA, 2004, 2006, 2009).

There are concerns from this study with regards to groundwater quality. Compounds related to surface agriculture such as nitrate leakage to greater depth and evidence of increasing salinity in groundwater (45-65 m bgs), threaten the quality of the water resource in this area. Increasing usage of recycled water for irrigated agriculture further increases the contamination potential of the deeper groundwater, hence its sustainability as a managed resource. The deterioration of groundwater to a depth beyond 30 m and the current position of the fresh-saltwater interface should generate more concern for the Groundwater Management Areas in this region. Protection of water quality in these areas should be considered in short- and long-term management strategies.

The declining water table in the shallow groundwater, which was initially feared following intensive abstraction between 1997 and 2007, had been successfully averted due to water restrictions and a seemingly return to wetter years since 2010. In addition, the use of recycled water eased the burden on groundwater. This is an example of informed management policy that needs to be upheld. Declining yields in bores would have become more evident or continued if rates of abstraction had not been checked through water restrictions in 2007. However, the rise in water table increases the chance of both lateral and horizontal hydraulic gradients that will promote flow between the shallow Delta and the deep Brighton Formation aquifers, which may have a negative impact on the quality of deeper groundwater in the future.

While localized areas of high EC around creeks (like Toolern and Skeleton creeks) exist within the basin, the connection to deeper groundwater through preferential flow paths and underlying geologic structures threatens the quality of the fresh groundwater resource. There are indications of downwards vertical mixing within the basin and lateral encroachment of sea water inlands. The area where mixing and contamination is highest is not only near the coast of Port Phillip Bay but further inland to about 10 km. The coastal part of the basin near Port Phillip Bay, under the current influence of landward hydraulic gradients, may suffer severe water quality deterioration in the future if long-term strategic management plans are not in place. The likelihood of short-term or long-term vertical flow through the Delta and basalts occurring is probably controlled by the degree to which lenses (intra-volcanic sands) are connected or interconnectivity of the aquifers underlying the Werribee Plains. Therefore, existence of higher salinity groundwater in the basin since regular water-quality monitoring began in the early-1980s and the movement of saline water from the coast to other parts of the basin should attract a more definitive management plan (e.g. less groundwater extraction) beyond the Deutgam Water Supply Protection Area. The fact that recharge occurs across the catchment indicates that protection of groundwater from surface contamination is dependent on the protection and management of large areas.

The groundwater samples from the Brighton Group are, on average, fresher and have lower $\delta^{18}O$ values than groundwater in the other aquifers, and this could be due to recharge during a previous wetter, cooler climate. This indicates that recharge to the Brighton Group is occurring, much more slowly than to the surface aquifer, i.e. connectivity to the Brighton Group is relatively restricted. This has management implications – this aquifer may be relatively protected from contamination.

With the economic viability of this region in terms of agricultural productivity, and given that irrigated agriculture in the area depends heavily on surface and ground water (and now with recycled water from the waste water treatment plants), the sustainability of groundwater usage should incorporate the long time-scales involved in recharge to the deeper aquifer unit and subsequently a longer term regional management plan. Strategic management plans become indispensable in coming years also with cyclic ‘drought and wetter year’ pattern of climate of this area, which can encourage increased irrigation and groundwater use. Such plans should include an initiative to facilitate the integration of isotope data into national water resource assessments. This initiative will assist in identifying gaps in existing hydrological information and in improving numerical models designed to simulate hydrological systems for resource management. Isotope techniques have an important role to play in providing fundamental information on water resources, as well as providing broader insight into aquifer characteristics.

CONCLUSION

The groundwater of Werribee basin appears well mixed, especially within the irrigation districts where there is conjunctive use of surface and ground waters with recycled water, due to intensive exploitation and the resulting drawdown and water table fluctuations. The LMWL developed for the study area is $\delta D = 6.98\ \delta^{18}O - 6.8$, whose slope is smaller than that of the GMWL, indicates the occurrence of secondary evaporation during rainfall. The effect of secondary evaporation is greatest for light precipitation; hence with the increase of rainfall intensity, the slope of the meteoric water lines becomes close to that of GMWL.

TDS and NO_3 contents are generally lower at the upper and mid-catchments compared to the lower catchment. This increase is likely due to a combination of sea water influence in addition to mineral dissolution and evapo-transpiration along groundwater flow paths. The highest concentrations are found in shallow groundwater samples of bores screened in the delta aquifer; however elevated concentrations also occur in deep bores (Brighton Group), particularly along the same flow path in the NW end of the Deutgam WSPA, indicating downward leakage of shallow water and vertical mixing. Due to the lack of natural nitrate in most geologic formations, the high NO_3 concentrations occur locally (up to 408 mg/L, median 70.6 mg/L) indicates contamination by fertilizers, animal wastes and/or effluents.

Despite the relatively similar stable isotope values, it was possible to estimate the main sources of origin for the waters of the basin. Measured groundwater $\delta^{18}O$ and δD values scatter around the LMWL indicating that groundwater is primarily of meteoric origin. The stable isotope data of groundwater, together with Cl and TDS, suggest a marine source of solutes for intermediate or deeper groundwaters. Both the $\delta^{18}O/Cl^-$ and Cl^-/Br^- ratios imply that evaporation (e.g. from shallow watertable) and transpiration (e.g. from the all-year-round cropping in the irrigation districts) during recharge represent the main processes controlling groundwater TDS contents across the basin. The shallow groundwater, particularly at the delta (delta aquifer) must have been derived from a combination of sources: the sea, recycled water and river delta water from surface irrigation and upstream groundwater from upper catchment area. The isotope ($\delta^{18}O$ - δ^2H) relationship in rainfall, surface and groundwater further reflects a degree of evaporation, common to most semi-arid climates. Groundwater often consists of a mixture of recharge from surface water (lakes or rivers) and local precipitation as demonstrated in this study. It is important to know the proportions of these recharge components in order to increase the sustainable supply of water, and to prevent pollution by infiltration of water from contaminated surface water sources.

The groundwater samples from the Brighton Group are, on average, fresher and have lower $\delta^{18}O$ values than groundwater in the other aquifers, and this was attributed to recharge during a previous wetter, cooler climate. This suggests that recharge to the deeper Brighton Group occurs slowly than to the surface aquifer, i.e. connectivity to the Brighton Group is relatively restricted. This has management implications – while the Brighton Group aquifer may be relatively protected from contamination, the rapid recharge across the delta has the most serious management implications, in terms of the almost instantaneous potential for contamination.

The conclusion that recharge occurs across the basin indicates that protection of groundwater from surface contamination is dependent on the protection and management of large areas. Protection of water quality in these areas should be a priority beyond the Deutgam Water Supply Protection Area and therefore generate shorter and long-term management strategies.

Reviewed by Dr Matthew Currell (School of Civil, Environmental & Chemical Engineering, RMIT University, Melbourne Australia), Assoc. Prof. John Webb (La Trobe University, Bundoora, Australia) and Dr Evan P. Dresel (Department of Environment and Primary Industries, Victoria, Australia)

REFERENCES

Adelana M. (2011). Changes in groundwater level dynamics in aquifer systems: implications for resource management in a semi-arid climate. In:

Muthukrishnavellaisamy Kumarasamy (eds.), Water Management / Book 1. InTech Open Access Publisher, Croatia.

Allison G.B., C.J. Barnes, M.W. Hughes, F.W.J Leaney. (1984). Effect of climate and vegetation on oxigen-18 and deuterium profiles in soils. In Isotope Hydrology 1983, IAEA Symposium 270, September 1983, Vienna, Austria, pp. 105–123.

Allinson, G., Salzman, S.A., Ii, H., McNamara, E., Taniguchi, M., Croatto, G., Kitching, M., Shelley, B. (2011). Understanding the connections between ground and surface water on the Werribee Plains. A report prepared for the Vision for Werribee Plains Initiative. DPI Queenscliff Centre.

Banks, EW, Simmons, CT, Love, AJ, & Shand, P. (2011). Assessing spatial and temporal connectivity between surface water and groundwater in a regional catchment: Implications for regional scale water quantity and quality, *J. Hydrol.*, 404 (1-2): 30–49.

Bates B, Zbigniew W, Shaohong W, Palutikof J. (2008). Climate Change and Water: Technical Paper VI. [Geneva, Switzerland]. UNT Digital Library. http://digital.library.unt.edu/ark:/67531/metadc11958/. Accessed October 25, 2013.

Bennetts DA, Webb JA, Stone DJM, Hill DM. (2006). Understanding the salinisation process of groundwater in an area of south-eastern Australia, using hydrochemical and isotopic evidence. *J. Hydrol.*, 323: 178-192.

Birch W. D. (ed). (2003). Geology of Victoria, Geological Society of Australia Special Publication 23.

Cartwright, I., Hall, S., Tweed, S., Leblanc, M. (2009). Geochemical and isotopic constraints on the interaction between saline lakes and groundwater in southeast Australia. *Hydrogeol. J.*, 17 (8): 1991–2004.

Cartwright, I; Morgenstern, U. (2012). Constraining groundwater recharge and the rate of geochemical processes using environmental isotopes and major ion geochemistry: Ovens Catchment, southeast Australia, *J. Hydrol.*, 475: 137–149.

CCSP (2008). The effects of climate change on agriculture, land resources, water resources, and biodiversity in the United States. A Report by the U.S. Climate Change Science Program and the Subcommittee on Global Change Research. U.S. Environmental Protection Agency, Washington, DC, USA.

Christopherson, N., Kjaernsrod, S., and Rhodhe, A. (1985). Preliminary evaluation of flow pattern in the Birkenes catchment using natural ^{18}O as a tracer. In: I. Johansson (Editor), Hydrological and Hydrogeochemical Mechanisms and Model Approaches to the Acidification of Ecological

Systems. IHP Workshop Proc., Uppsala. Nord. Hydrol. Program Rep. No. 10: 29-40.

Clark, I, Fritz, P. (1997). Environmental Isotopes in Hydrogeology. New York: Lewis Publishers.

Coombes PJ, Colegate M, Bairstow J, McGrath A. (2010). Integrated stormwater management at the Werribee Employment Precinct. Werribee City Infrastructure Planning Integrated Water Cycle Management, Planning for the productive use of storm water. Bonacci Water Pty Ltd. 50 Hoddle Street Abbotsford 3067 Australia.

Craig, H. 1961. Isotopic variations in meteoric waters. *Science,* 133:1702–1703.

Crosbie RS, Morrow D, Cresswell RG, Leaney FW, Lamontagne S and Lefournour M. 2012. New insights into the chemical and isotopic composition of rainfall across Australia. CSIRO Water for a Healthy Country Flagship, Australia.

Currell et al. (2010). Environmental isotopes in groundwater to understand the response of a vulnerable coastal aquifer. *J. Hydrol.,* 385: 216-229.

Currell, M.J., Cendon, D, Cheng X. (2013). Analysis of environmental isotopes in groundwater to understand the response of a vulnerable coastal aquifer to pumping: Western Port Basin, south-eastern Australia. *Hydrogeol. J.,* doi:10.1007/s10040-013-1017-9.

Dahlhaus, P.G., Heislers, D.S., Brewin, D., Leonard, J.L., Dyson, P.R., Cherry, D.P. (2004). Port Phillip and Westernport Groundwater Flow Systems. Port Phillip and Westernport Catchment Management Authority, Melbourne, Victoria.

Dincer, T, Payne, B.R., Florkowski, T., Martinec, J. and Tongiorgi, E. (1970). Snowmelt runoff from runoff of tritium and oxygen-18. *Water Resour. Res.*, 18: 977-996.

DSE (2008). Climate change in Port Phillip and Western Port. The Victorian climate adaptation program.

Epstein, S., Mayeda, T.K. (1953). Variations of the $^{18}O/^{16}O$ ratio in natural water. *Geochim. Cosmochim. Acta,* 4:213–220.

GeoAust (2009). Preliminary geotechnical assessment of the Werribee Employment Precinct

Gat, J.R., Mook, W.G., Meijer, A.J. (2001). Atmospheric water. In: Mook WG. (ed) Environmental isotopes in the hydrological cycle: principles and applications, vol 2. IHP-V, Technical document no. 39. UNESCO, Paris.

Gourcy, L.L., Groening, M., Aggarwal, P.K. (2005). Stable oxygen and hydrogen isotopes in precipitation. In: Aggarwal, P.K., Gat, J.R.,

Froehlich, K.F.O. (eds) Isotopes in the water cycle: past, present and future of a developing science. Elsevier, Amsterdam, pp 39-51.

Hare, A.G., Cas R.A.F., Musgrave, R., Phillips D. (2005). Magnetic and chemical stratigraphy for the Werribee Plains basaltic lava flow-field, Newer Volcanics Province, southeast Australia: implications for eruption frequency. *Australian Journal of Earth Sciences,* 52: 41–57.

Harrington, G.A., Cook, P.G., Herczeg, A.L. (2002). Spatial and temporal variability of groundwater recharge in Central Australia: A tracer approach. *Ground Water,* 40(5):518-528.

Heaton, T.H.E. 1986. Isotopic studies of nitrogen pollution in the hydrosphere and atmosphere: a review. *Chem. Geol.,* 59: 87–109.

Herczeg, A.L., Dogramaci S.S, Leaney FW. (2001). Origin of dissolved salts in a large, semi-arid groundwater system: Murray Basin Australia. *Marine Freshwater Res.,* 52: 41–52.

Holdgate, G.R.; Gallagher, S. J. & Wallace, M. W. (2002). Tertiary coal geology and stratigraphy of the Port Phillip Basin, Victoria, *Australian Journal of Earth Sciences,* 49:437–453.

Hughes, C.E., Crawford J. (2012). A new precipitation weighted method for determining the meteoric water line for hydrological applications demonstrated using Australian and global GNIP data. *J. Hydrol.,* 464–465:344–351.

IAEA/WMO. (2006). Global Network of Isotopes In Precipitation. The GNIP Database. Accessible at http://www.iaea.org/water.

Ii, H., Satoh, A., Taniguchi, M., Kitching, M., Croatto, G., Shelley, B., Allinson G. (2012). Groundwater Contamination Due to Irrigation of Treated Sewage Effluent in the Werribee Delta. *Int. J. of GEOMATE,* 3, No. 1 (Sl. No. 5): 332-338.

Ingraham, N.L. (1998). Isotopic variation in precipitation. In: Kendall C, McDonnel JF (eds) Isotope tracers in catchment hydrology. Elsevier, Amsterdam.

Kennedy, V.C., Kendall, C., Zellweger, G.W., Wyerman, T.A. and Avanzino, R.J., (1986). Determination of the components of stormflow using chemistry and environmental isotopes, Mattole River basin, *California. J. Hydrol.,* 84: 107-140.

LeBlanc et al, (2009). Millenium Drought, a-documented hydrological event in the – Water Resources Research, issue 45

Leonard, J. (2008). Inquiry into Melbourne's Future Water Supply. Submission to the Environment and Natural Resources Committee.

Submission No# 28. John Leonard Consulting Services, Groundwater and Environmental Consultants. 19th August 2008.

Lowe L.D., Webb, J.A., Nathan, R.J., Etchells, T., Malano, H.M. (2009). Evaporation from water supply reservoirs: An assessment of uncertainty. *J. Hydrol.,* 376(1–2): 261-274.

Mazor, E. (2004). Chemical and isotopic groundwater hydrology. Dekker, New York.

McCarthy, K.A., McFarland, W.D., Wilkinson, J.M., White, L.D. (1992). The dynamic relationship between groundwater and the Columbia River: Using deuterium and Oxygen-18 as tracers. *Journal of Hydrol.,* 135(1-4): 1-12.

Mook, W.G. (2006). Introduction to isotope hydrology: stable and radioactive isotopes of hydrogen, oxygen and carbon. Taylor & Francis, London.

Mook, W.G, Groenveld, D.J., Brown, A.E., and Van Ganswijk, A.J. (1974). Analysis of a runoff hydrograph by means of natural isotopes. Proc. IAEA Symp. Isot. Tech. Groundwater Hydrol., Vienna, pp. 145-153.

O'Neil, J.R., Adami, L.H., Epstein, S. (1975). Revised value for the ^{18}O fractionation factor between CO_2 and water at 25°C. *J. Res.,* 3:623–624.

Pittock, J. (2011). National climate change policies and sustainable water management: conflicts and synergies. *Ecology and Society,* 16(2): 25. [online] URL: http://www.ecologyandsociety.org/vol16/iss2/art25/.

Rodhe, A. (1984). Groundwater contribution to streamflow in Swedish forested till soil as estimated by oxygen-18. Proc. Syrup. Isot. Hydrol., 1983, IAEA, Vienna pp. 55-66.

Rozanski, K., Araguas-Araguas, L., Gonfiantini, R. (1993). Isotopic patterns in modern global precipitation. In: Swart, P.K., Lohmann, K.C., McKenzie, J., Savin, S. (Eds.), Climate Change in Continental Isotopic Records. AGU Geophysical Monograph Series. Washington, DC, pp. 1–36.

Saha, D., Dwivedi, S.N., Goutam, K.R., Reddy, D.V. (2013). Isotope-based investigation on the groundwater flow and recharge mechanism in a hard-rock aquifer system: the case of Ranchi urban area, India. *Hydrogeol,* 21: 1101-1115.

Sklash, M.K.; Farvolden, R.N. (1982). The use of environmental isotopes in the study of high-runoff events in streams. In: E.C. Perry and C.W. Montgomery (Editors), Isotope Studies of Hydrologic Processes. N. Illinois University Press, Dekalb, Ill., pp. 65-73.

Sklash, M.K.; Farvolden, R.N.; Fritz, P., (1976). A conceptual model of watershed response to rainfall, developed through the use of oxygen-18 as a natural tracer. *Can. J. Earth Sci.*, 13: 271-283.

SKM (2004). Deutgam groundwater Investigation. Summary report on Groundwater modelling. Sinclair Knight Mercz, August 2004.

SKM (2005). Werribee Irrigation District groundwater investigations. Sinclair Knight Mercz, August 2005.

SRW (2004). Annual Report, Southern Rural Water File Download: Annual_reports/Annual_Report_2004.pdf

SRW (2006). Annual Report, Southern Rural Water File Download: Annual_reports/Annual_Report_2006.pdf

SRW (2009). Western Irrigation Futures Atlas. Southern Rural Water, June 2009.

Simpson, H. J., Herczeg A. L. (1991). Stable Isotopes as an Indicator of Evaporation in the River Murray, Australia, *Water Resour. Res.*, 27(8): 1925–1935.

Turner, J.V., Macpherson, D.K., Stokes, RA. (1987). The mechanisms of catchment flow processes using natural variations in deuterium and oxygen-18. *J. Hydrol.,* 94: 143-162.

Visualizing Victoria's Groundwater (2011). The Victorian Water Measurement Information System (WMIS) Victoria Department of Sustainability and Environment Water Resources Database http://www.vvg.org.au/cb_pages /wmis.php,

Vrbka, P., Jacob, H., Froehlich, K., Salih, M. (1993). Identification of groundwater recharge sources in northern Sudan by environmental isotopes. *J. Environ Hydrol.,* 1(4):98-125.

Wei et al. (2013). Dynamics of groundwater-derived nitrate and nitrous oxide in a tidal estuary from radon mass balance modeling. *Limnol. Oceanogr.,* 58(5): 1689–1706.

Yao, Z.J., Lin, J., Huang, H.Q., Song, Z.F., Dong, X.H., Liu, X. (2009). Characteristics of isotope in precipitation, river water and lake water in the Manasarovar Basin of Qinghai: Tibet Plateau. *Environ Geol.*, doi: 10.1007/s00254-008-1324-y

Yin, L.H., Hou, G.C., Su, X.S., Wang, D., Dong, J.Q., Hao, Y.H., Wang, X.Y. (2011). Isotopes (δD and δ18O) in precipitation, groundwater and surface water in the Ordos Plateau, China: implications with respect to groundwater recharge and circulation. *Hydrogeol. J.*, doi:10.1007/s10040-010-0671-4

In: Groundwater
Editor: Segun Michael Adelana
ISBN: 978-1-63321-759-1

Chapter 8

Groundwater Streams Running over Clay on the Talus Slope of the Niagara Escarpment (Canada): An Overview with Ecological Implications

H. J. Swatland*
University of Guelph, Guelph, Canada

Abstract

Cold groundwater in a river bed creates suitable conditions for the spawning of brook trout (*Salvelinus fontinalis*). In some locations along the Niagara Escarpment, groundwater discharged from a primary aquifer may run, exposed to the atmosphere, down a slope of clay into the river. At other locations, the flow above clay may be within the overburden and only exposed at odd sites, such as the stream vent described in this chapter. In exposed streams, travertine deposited by the degassing of calcium-rich groundwater after passing over clay is darker than upstream deposits. Spectrocolorimetry of travertine, clays and shale, and electrical impedance of clays were used to gather supporting information. Ecological sampling showed a rich river fauna in the study area, mostly associated with pristine water.

* Corresponding author: e-mail: swatland@uoguelph.ca; Telephone: (519) 821-7513.

Keywords: Niagara Escarpment, Travertine, Clay, Spectrocolorimetry, Electrical impedance, Spring vent, Cation exchange

INTRODUCTION

The Niagara Escarpment is a major geological feature in North America – a long cuesta arching through the Great Lakes Basin and forming the background to Niagara Falls. In the study area (Figure 1), groundwater is discharged from the Niagara Escarpment at two levels, at an upper level below the dolostone of the Amabel Formation and at a lower level above the red and green shales of the Queenston Formation (Brogly et al. 1998; Rodie and Post, 2009). The groundwater may flow in several ways over clays formed from weathered shale, flowing either in waterfalls, in broad seepages, or in streams, which may be fully exposed to the atmosphere or subterranean and running between talus rocks and within the overburden.

As first observed by Sewell in 1837, groundwater discharged from the Niagara Escarpment may deposit travertine (calcareous tufa) around waterfalls. The terminology on this topic is still contentious, tufa *versus* travertine, as in Ford and Pedley (1996) *versus* Pentecost (2005), but the chemistry is fairly straightforward. Where the groundwater contains $Ca(HCO_3)_2$, it may deposit $CaCO_3$ when CO_2 is degassed (Swatland, 2010), especially in the presence of a microbial biofilm (Rogerson et al. 2008). On the Niagara Escarpment, where travertine typically gets deposited on a locally dominant moss such as *Didymodon tophaceus,* removal of CO_2 by photosynthesis does not seem too important because travertine is deposited deeply in the bed of moss and moss leaves cast strong shadows (Swatland, 2011).

As in other geographical locations, travertine formed in this way typically forms a yellowish-white encrustation (Emig 1917; Pentecost 2005). But, if $CaCO_3$ is still being deposited downstream after passing over clay, a variety of colored epilithic deposits may be formed, with or without the addition of chlorophyll from encrusting algae (Bergey and Weaver, 2004). This epilithic deposit sometimes forms a continuous layer to stabilize silt, rock fragments and plant material and, thus, is of interest in several fields of study such as taphonomy (Diéguez et al., 2009; McKee, 2010), diagenesis (Love and Chafetz, 1988; Riding, 2000) and ecology (Habdija et al., 2004; Viles et al., 2008; Compson et al., 2009).

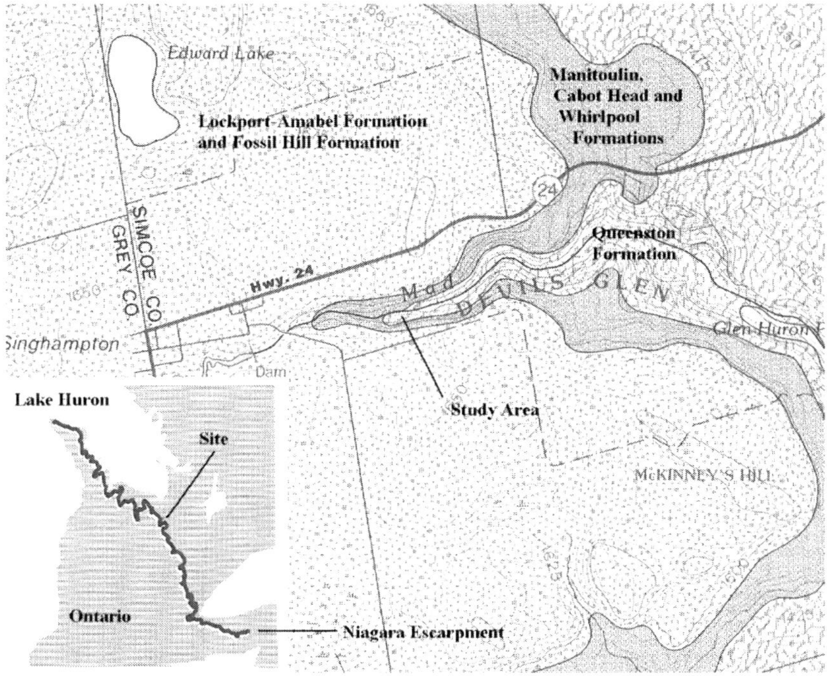

Figure 1. Location of the study area near Singhampton, Ontario, on the Niagara Escarpment (adapted from Tovell 1992). At the study area, the descending stratigraphic sequence is Silurian gray dolostone of the Amabel Formation, Silurian crystalline dolostone of the Manitoulin Formation, and Ordovician red shale of the Queenston Formation (Telford 1976).

This chapter starts with a geomorphological description of a somewhat unusual feature of a groundwater stream – a spring vent. Spring vent hydrology is then compared with adjacent exposed streams and river water. Spectrocolorimetry is used to gather support for the working hypothesis that cation exchange with clays might darken travertine downstream from clay. The possible ecological importance of exposed *versus* subsurface groundwater streams is discussed.

SPRING VENT

On the Niagara Escarpment north of Niagara Falls, the Mad River runs through a preglacial notch in the cuesta, then joins the Nottawasaga River to empty eastwards into Georgian Bay of Lake Huron (Chapman and Putnam,

1973). Here the full stratigraphic column of the cuesta has been eroded so that the top layer in the headwaters of the Mad River Valley is the Amabel Formation of the Lower Silurian (Tovell, 1992). As the river cuts eastwards down the cuesta, the lower levels of the Clinton and Cataract Groups (Middle and Lower Silurian, respectively) are exposed above the Queenston Formation (Upper Ordovician).

Groundwater is discharged from the Niagara Escarpment at two exposure points, at an upper point below the dolostone of the Amabel Formation and at a lower point above the red and green shales of the Queenston Formation (Rodie and Post, 2009) and then flows in waterfalls, broad seepages and surface streams. The subterranean streams running over clays but between talus rocks and within the overburden are not well known and raise some questions.

The northern talus slope of the Mad River Valley at Universal Transverse Mercator 561395E, 4911204N contains a small structure reminiscent of a karst window - a depression revealing part of an underground stream flowing across its floor (Figure 2).

Figure 2. Spring vent. At this time (November, 2012), it was almost 50 cm across and several centimetres deep. Water entered at approximately 0.1 m/s at a point near the top middle of the picture and drained across the floor of the vent. The turbulence at the water surface is visible. The vent is noisy and remains open in deep snow.

But being on a talus slope dominated by blocks of dolostone, a designation as a karst window is misleading, because the surrounding structure is not eroded rock and the water level is not depressed relative to the surrounding ground level. From the comprehensive glossary of karst terminology compiled by Copeland (2003), the most suitable term is a spring vent – an opening that concentrates groundwater discharge at the Earth's surface. Other terms such as estavelle, swallet and ponor are inappropriate because they are normally applied to structures down which water sinks at certain times or which are intermittent or reversible in flow direction.

HYDROLOGY

The area surrounding the spring vent is a steep talus slope beneath the Niagara Escarpment with compacted leaf litter around living trees (*Thuja occidentalis,* Eastern White Cedar, and *Betula papyrifera,* White Birch). To find the characteristics of adjacent groundwater, the cuesta was hiked eastwards from the spring vent to locate five sites where groundwater was discharged directly from the primary aquifer at the base of the cuesta (measurements at five sites on different hikes to give n = 19). To find the characteristics of surface streams entering the river, the Mad River was hiked eastwards (n = 72 for three streams repeatedly sampled).

Water temperature, *p*H and conductivity were measured with a waterproof meter calibrated before each set of measurements (PC300, Oakton Instruments, Vernon Hills, IL). The *p*H electrode was cleaned regularly with pepsin and HCl (Oakton 653-06) and stored in an appropriate medium (Oakton 653-04). Measurements were made 20 seconds after the ready indicator appeared.

The pH of water in the spring vent did not differ ($P > 0.05$) from the pH of groundwater directly discharged from nearby aquifers but was lower ($P < 0.001$) than the pH of water in streams where they discharged into the Mad River (Table 1). Degassing of CO_2 may cause an increase in the pH of groundwater after it is discharged from its primary aquifer (Choi et al., 1998), so the pH data in Table 1 indicate the spring vent is close to the primary aquifer or, at least, that there is minimal degassing in a secondary aquifer. But the conductivity data are more complex.

One expects conductivity to decrease as consequence of degassing and the precipitation of calcite, thus decreasing Ca^{2+} concentration (Chen et al., 2004);

which agrees with groundwater at its primary aquifer having a higher conductivity (Table 1, $P < 0.01$) than at the discharge of streams into the river.

Table 1. Hydrology at the spring vent relative to exposed streams on the same slope

Measurement	Exposed streams at source	Spring vent	Exposed streams at river
Acidity, pH	7.45 ± 0.25	7.51 ± 0.20	$8.06 \pm 0.17^{*}$
Conductivity, μS	$637 \pm 114^{*}$	466 ± 74	$573 \pm 45^{*}$

But what might cause conductivity to be lower (Table 1, $P < 0.001$) in the spring vent than at either nearby groundwater sources or discharge into the river? In an exposed stream on clay, the general pattern is for conductivity to decrease as pH increases (Swatland, 2012). Here in the spring vent we have a low conductivity with a minimal increase in pH. If low pH indicates acidic water which we would expect to have high $Ca(HCO_3)_2$, what is the cause of the low conductivity? Possibly, the source water may have picked up CO_2 from the soil but, in a short dwell time as groundwater, not dissolved much $CaCO_3$. However, sampling error is the most likely explanation in this case. Conductivity data from the spring vent are shown in relation to a nearby exposed spring in Figure 3.

The exposed spring was sampled rapidly at measured intervals from source to the river at an air temperature of 13.5 °C, through which water temperature increased from 6.9 to 8.0 °C giving a strong correlation of pH with conductivity ($r = -0.78$, $P < 0.001$, line in Figure 3). The solid squares in Figure 3 show the spring vent data collected over a period of two years with air temperatures ranging from -9 to 21°C, and water temperatures ranging from 2.6 to 8.1°C. This looks like sampling error caused by temperature.

Groundwater entering the Mad River from an exposed stream was detectable by pH, as in the example shown in Figure 4 measured at low water (October 2012). Groundwater at pH 7.55 and 8.2 °C entering river water at pH 8.16 and 10.4 °C was measured across two downstream transects across the river; the first at 2 m from the entry point (Figure 4A), and the second at 18.3 m from the entry point (Figure 4B). At the first transect the mean depth was 0.18 ± 0.08 m with a river width of 12.4 m and, at the second transect, the mean depth was 0.50 ± 0.08 m with a river width of 20.7 m.

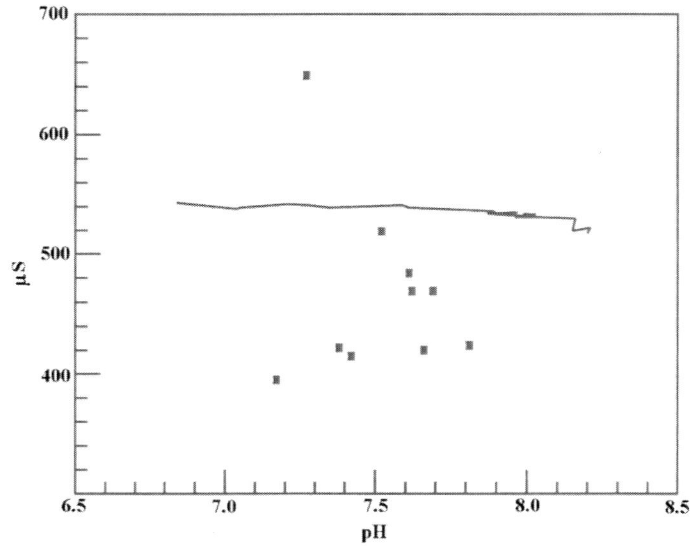

Figure 3. Two-year relationship of pH with conductivity (µS) at the spring vent (■) relative to a line showing this relationship in an exposed groundwater stream without seasonal variation ($r = -0.78$, $P < 0.001$).

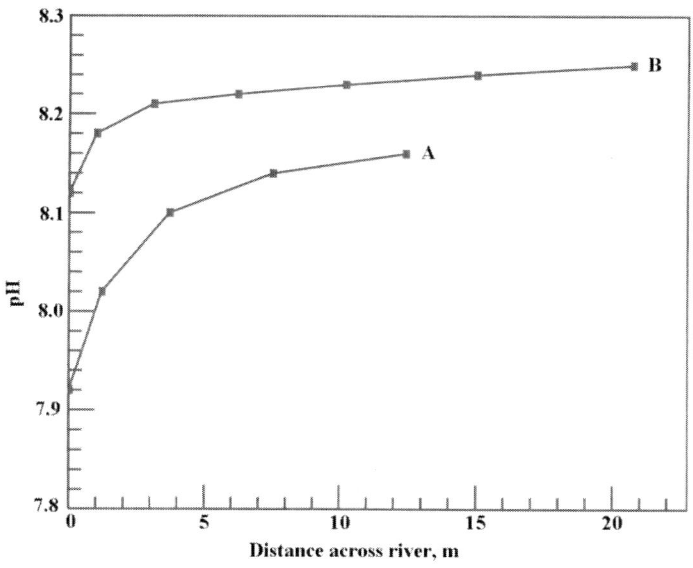

Figure 4. Effect of groundwater entering the river from a lateral waterfall at distance 0 m across the river. Transects were 2 m (A) and 18.3 m (B) downstream from the entry point.

River water pH was lower ($P < 0.05$) on the side where groundwater entered relative to the opposite side. No differences for conductivity or temperature across the river were detected.

PRINCIPLES OF SPECTROCOLORIMETRY

The colors of shales are essential in stratigraphic correlation and depositional analysis (Potter et al., 1980), although the mineral basis of color in shale is sometimes contentious (Ferretti et al. 2012).

The stepwise Munsell rock-color chart (Munsell, 2009) is useful for communication, but requires subjective matching of rocks to Munsell color chips. Objective colorimetry has proved useful in the study of clays (Nagano et al., 1992; Scheinost et al., 1998), so fiber-optic spectrophotometry was used here to gather data for weighted-ordinate estimates of CIE (Commission Internationale de l'Éclairage) chromaticity coordinates (Billmeyer and Saltzman, 1981). Reflectance values across the visible spectrum (400 to 700 nm in 10 nm steps with a 10 nm band-pass) were weighted by the sensitivity of the eye to those wavelengths. In colorimetry, the general type of color implicit in widely used terms such as red or green shale, is called a hue. If weathering of a red shale increases light scattering, then it may appear brighter, yet the hue may be unchanged. But, in general, the property of color which changes in this example is the intensity of the color, and this is called the chroma. In the CIE method, the primary hues, red, yellow, green and blue, are added or subtracted from each other to match any color. By a mathematical manipulation, it is possible to specify both hue and chroma in the CIE system by means of a single pair of chromaticity coordinates called x and y (Billmeyer and Saltzman, 1981)). Changes in hue follow a circumferential pattern in the CIE diagram, while changes in chroma radiate from the central position of white (although the coordinates of achromatic white depend on the emission spectrum of the illuminator). Luminosity (here called $Y\%$) may be specified by a third coordinate relative to the plane of the chromaticity coordinates.

SPECTROCOLORIMETRY APPARATUS

The spectrocolorimeter was based on a Zeiss Universal microscope (Carl Zeiss, D-7082 Oberkochen, Germany) connected to rock samples by fiberoptics as described by Swatland (1998). The main controller was a personal computer running HP BASIC (Hewlett-Packard, Palo Alto, CA). The photomultiplier was a side-window Hamamatsu (1126 Ichino-Cho, Hamamatsu City, Japan) HTV R 928 with S-20 characteristics. A grating monochromator with stray-light filters was mounted under the photomultiplier and scanned from 400 to 700 nm in steps of 10 nm with a 10 nm band-pass. Illumination was from a 100 W halogen source with a stabilized power supply directed through a solenoid shutter. The bifurcated fiber-optic probe (WW 100-1, Guided Wave, Eldorado Hills, California) was mounted with the illumination branch in place of a vertical epicondenser on the microscope. The standard for diffuse white reflectance was Teflon tape (politetrafluoroethylene; Weidner & Hsia, 1981) wrapped around microscope slides, with attention given to the clearance between the standard and the common trunk of the light guide which was adjusted to give a maximum at 550 nm. Light from the measuring branch of the bifurcated light guide was directed into the microscope spectrophotometer by a swing-in mirror in the optical axis of the microscope.

Color technology is important industrially and is well supported by a wide range of methods for colorimetry which makes the choice of reporting methods somewhat arbitrary. But, the CIE (Commission Internationale de l'Éclairage) system using the 1931 weighted ordinates is the obvious choice (Billmeyer and Saltzman, 1981). Although the chromaticity coordinates cannot be claimed as true CIE values because of the nature of the emission spectrum used for their measurement, the results reported here gave a close match to hue and chroma on a large-scale CIE chart (Zeiss Pl 50-661-e). Spectra were tested by a t-test at each wavelength (Steel and Torrie, 1980).

UPSTREAM VERSUS DOWNSTREAM TRAVERTINE

The reflectance spectra of travertine above and below stream exposure to clay are shown in Figure 5. All differences were significant, $P < 0.01$, and from 530 to 690 nm differences were significant, $P < 0.005$ ($n = 10$ for each type, 1-tailed t-test). Was groundwater changed by passing over clay, or was this color difference in travertine caused by the mechanism of deposition – splashing *versus* aquatic laminar flow?

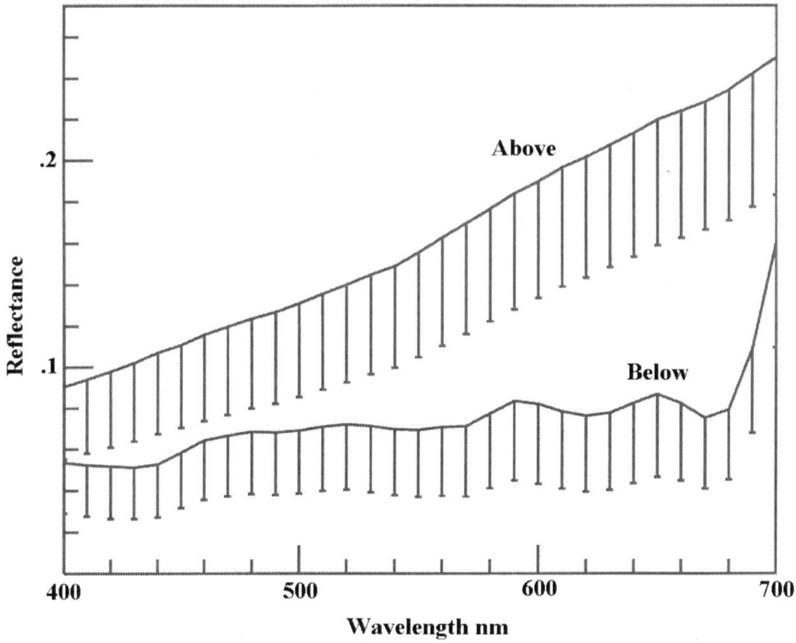

Figure 5. Reflectance spectra of travertine above and below stream exposure to clay. Error bars show mean – SD ($n = 10$).

Spectrocolorimetry of Clay

Queenston Formation shale typically gives rise to a generally agreed red clay when fully oxidized, and to a reduced form variously described as gray or green (Vos, 1975) although, judging from the large number of local names for blue springs emanating near the Queenston Formation, early settlers in Ontario along the Niagara Escarpment thought the clay was blue. In the field, the subjective impression of gray or blue is strongly affected by whether the sky is gray with stratus clouds or completely clear blue with nothing but Rayleigh scattering. Similarly, whether or not oxidation-reduction status was at the time of geological deposition, or from later exposure to acidic groundwater is debatable. At the study sites, the oxidized clay was grayish-red (Munsell 5R 4/2) while the reduced clay was greenish-gray (Munsell 5G 6/1). Reflectance spectra for grayish-red and greenish-gray clays are shown in Figure 6. Differences between mean spectra were significant ($P < 0.001$, $n = 10$ of each type, 2-tailed t-test), as were chromaticity coordinates (for grayish-red, $x =$

0.3867 ± 0.0079, y = 0.3496 ± 0.0079, $Y\%$ = 7.4375 ± 0.6942; and for greenish-gray, x = 0.3283 ±0.0008, y = 0.3415 ± 0.0009, $Y\%$ = 31.5986 ± 1.5038; $P < 0.005$).

Readers who might check these chromaticity coordinates against a CIE color chart may find all the data intersect in an achromatic or colorless zone where printing ink fails to render various hues of gray because the 3-dimensional CIE space has been reduced to a 2-dimensional plane by collapsing luminosity ($Y\%$) in the third dimension. This prompted a check on the instrumentation, measuring interference filters of various hues which were correctly located to the periphery of the CIE chart. This also highlights a weakness in using CIE colorimetry for rocks with a minimal hue because, in many cases, a reader will have difficulty in finding a published color chip for subjective appraisal. This is where the Munsell color chips for rocks are most useful.

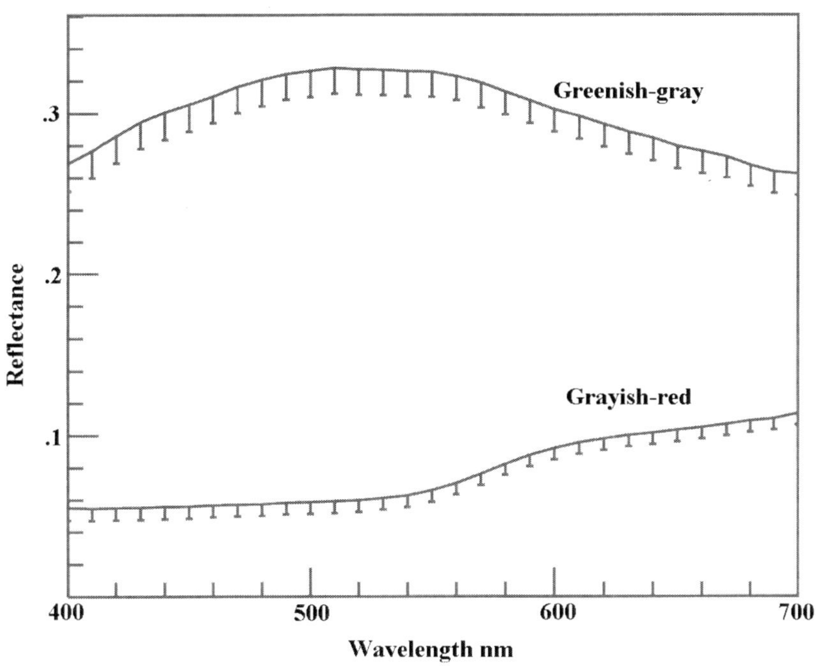

Figure 6. Fiber-optic reflectance spectra of greenish-grey and grayish-red clays derived from Queenston Formation shale (error bars show mean – SD for n = 10).

SPECTROCOLORIMETRY OF SOURCE SHALE

When dry Queenston shales giving rise to oxidized and reduced clays were measured, the spectra were comparable with those of the clays (Figure 7). The oxidized shale was light brown (Munsell 5YR 6/4) and the reduced shale was yellowish-gray (Munsell 5Y 7/2). Overall reflectance was higher for reduced shale relative to oxidized shale ($P < 0.005$ from 400 to 410 nm, and $P < 0.0005$ from 420 to 700 nm), with CIE values for reduced shale ($x = 0.3615 \pm 0.0023$, $y = 0.3638$, $Y\% = 14.4797 \pm 1.4839$) and oxidized shale ($x = 0.3671 \pm 0.0021$, $y = 0.3449 \pm 0.0015$, $Y\% = 8.2270 \pm 2.1549$) differing significantly ($P < 0.0005$, $n = 10$ of each). Possibly, these differences originated from redox conditions at the time of deposition, but one cannot be sure (Wilde and Quinby-Hunt, 2010).

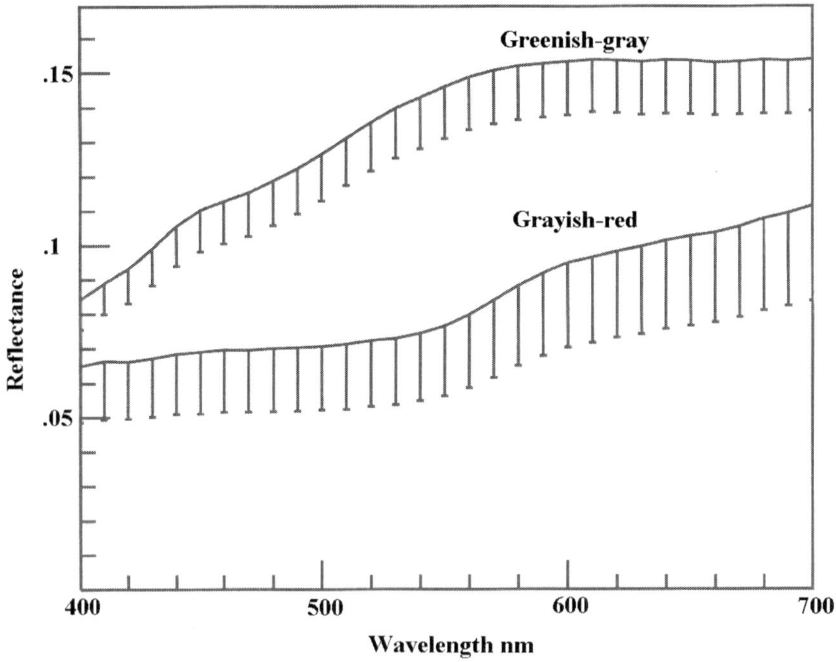

Figure 7. Fiber-optic reflectance spectra of Queenston Formation shales giving rise to greenish-grey and grayish-red clays (error bars show mean – SD for $n = 10$).

Electrical Impedance of Clay

If we follow a working hypothesis that cation exchange (Eckstein et al., 2007) in groundwater running over clay might explain differences in travertine color above and below the clay, we may ask how different colors of clay (grayish-red *versus* greenish-gray) might differ in this regard. Exposed groundwater streams at the study area passed over complex patterns of different colored clays. As explained by Mehran and Arulanandan (1977), electrical conductivity of colloid-water-electrolyte systems increases with the frequency of the test current (conductivity dispersion). In clays, conductivity dispersion increases with a reduction in water content, electrolyte concentration and cation-exchange capability. Thus, high conductivity (low resistance) may be an indication of cation-exchange capability.

Stiff, dry clay samples were impaled with parallel needles (gauge 18, length 4 cm, axial separation 17 mm) and measured with a Hewlett Packard 4262A LCR meter using a parallel circuit mode at 1 kHz. Greenish-gray clay had a much lower ($P < 0.001$) resistance (328.2 ± 34.0 Ω) than grayish-red clay (820.6 ± 158.8 Ω). Clays also may exhibit an extremely high dielectric constant (Arulanandan and Mitchell, 1968). Greenish-gray clay had a much higher ($P < 0.001$) capacitance (24.6 ± 4.7 nF) than grayish-red clay (5.8 ± 0.8 nF). Thus, greenish-gray clay may have been more likely than grayish-red clay to participate in cation exchange with groundwater.

Ecological Implications

Ricker (1931) started his pioneering ecological work on *Salvelinus fontinalis*, the brook or speckled trout, in Ontario's Mad River, so this river has a special importance in the history of fisheries research. Cold groundwater percolating through gravel creates optimal conditions for spawning (Webster and Eiriksdottir, 1976; Witzel and Maccrimmon, 1983; Curry et al., 1997). Groundwater streams running over talus slopes and into the Mad River are easily accessible for study, and they certainly contribute water at a fairly constant temperature throughout the year, despite atmospheric warming in the summer (Swatland, 2012). But this may be missing the special importance of cold groundwater entering directly into the bed of the river – something that is difficult to measure directly.

Over the Niagara Escarpment, the water in the Mad River is deep, fast and dropping over a dangerous terrain. If damage to a Biosphere Reserve was of no concern, one might dig down through the talus slope to sample underground streams but, of course, this is not an option. Hence, the attention given to the spring vent described in this chapter – because it may give us an insight into the groundwater entering unseen into the gravel beds of spawning trout. Quite possibly, there may be many other spring vents on the slopes below the Niagara Escarpment awaiting discovery and investigation.

From the local effect of adding groundwater to river water from a waterfall on the side of the river (Figure 4), one may argue that groundwater entering directly to the bed of the river would have an even stronger effect, for two reasons.

- Firstly, there might be slower turbulent mixing from a seepage through the river bed relative to a lateral waterfall.
- Secondly, cold groundwater might remain longer at the bottom of the river than when splashed in.

The hyporheic zone where groundwater may enter a river bed is now widely recognized as an important ecosystem with follow-on effects to the whole river (Boulton et al., 2010). Table 2, gives a *prima facie* identification of animals recovered from the study area. Identification to the species level was difficult but there was no doubt of the richness of the fauna in the river–many animals were indicative of pristine water conditions.

CONCLUSION

On talus slopes below the Niagara Escarpment, the transition zone from aquifer groundwater to surface water may be complicated by the presence of clay. Geomorphologically, on a small scale, clay may determine whether groundwater descends in waterfalls, seepages or underground streams. The resulting differences in atmospheric exposure mediated via differences in heat exchange and degassing may have an ecological impact on aquatic fauna. Cation exchange as groundwater streams pass over clay is a possibility, leading to darkening of travertine deposits.

Table 2. Aquatic fauna in the study area

River bed			
Protozoa	Ciliophora		*Frontonia* sp.
			Vorticella sp.
Rotifera	Bdelloidea		*Rotaria* sp.
Platyhelminthes	Tricladia		*Planaria* sp.
Annelida	Clitellata		*Placobdella* sp.
Insecta	Megaloptera		*Neohermes* sp.
			Corydalus cornutus
	Trichoptera		*Macronemum zebratum*
			Rhyacophila lobifera
			Hydropsyche sp.
Ephemeroptera	*Acerpenna* sp.		
	Leucrocuta sp.		
	Ephemerella sp.		
	Diphetor sp.		
	Heptagenia sp.		
Hemiptera	*Rhagovelia* sp.		
Coleoptera	*Graphoderus liberus*		
	Stenelmis sp.		
Plecoptera	*Acroneuria evoluta*		
	Capnia vernalis		
Odonata	*Libellula pulchella*		
Arachnida	Acari	*Hydrachnidia* sp.	
Crustacea	Decapoda	*Cambarus bartonii*	
		Oronectes propinquus	
	Cladocera	*Simocephalus* sp.	
		Cypria sp.	
		Allonela sp.	
Vertebrata	Cypriniformes	*Catastomus commersonii*	
Exposed streams			
Protozoa	Ciliophora		*Epalxella* sp.
			Spirostomum sp.
			Paramecium sp.
			Vorticella sp.
	Euglenophycota *Anisonema* sp.		
Rotifera	Eurotatoria		*Trichotria* sp.
			Dicranophorus sp.
Mollusca	Gastropoda		*Bakerilymnaea dalli*
Insecta	Diptera		*Euparyphus* sp.
Crustacea	Amphipoda		*Hyalella* sp.
Vertebrata	Amphibia		*Plethodon cinereus*

References

Al Nasser, W.N.; Al Salhali, F.H. (2013). Scaling and aggregation kinetics determination of calcium carbonate using inline technique. *Chem. Eng. Sci.* 86:70-77.

Arulanandan, K.; Mitchell, J.K. (1968). Low frequency dielectric dispersion of clay-water-electrolyte sysytems. *Clays Clay Minerals* 16: 337-351.

Bergey, E.A.; Weaver, J.E. (2004). The influence of crevice size on the protection of epilithic algae from grazers. *Freshwater Biol.* 49: 1014-1025.

Billmeyer, F.W.; Saltzman, M. (1981). *Principles of Color Technology*. Wiley, NY. 240 pp.

Brogly, P.J.; Martini, I.P.; Middleton, G.V. (1998). The Queenston Formation shale-dominated mixed terrigenous-carbonate deposits of Upper Ordovician, semiarid, muddy shores in Ontario, Canada. *Can. J. Earth Sci.* 35: 702-719.

Boulton, A.J.; Datry, T.; Kasahara, T.; Mutz, M.; Stanford, J.A. (2010). Ecology and management of the hyporheic zone: stream-groundwater interactions of running waters and their floodplains. *J. N. Am. Benthol. Soc.* 29: 26-40.

Chapman, L.J.; Putnam, D.F. (1973). *The Physiography of Southern Ontario*. University of Toronto Press, Toronto, ON. p. 179.

Chen, J.; Zhang, D.D.; Wang, S.; Xiao, T.; Huang, R. (2004). Factors controlling tufa deposition in natural waters at waterfall sites. *Sediment. Geol.* 166: 353-366.

Choi, J.; Hulseapple, S.M.; Conklin, M.H.; Harvey, J.W. (1998). Modeling CO_2 degassing and pH in a stream-aquifer system. *J. Hydrol.* 209: 297-310.

Compson, Z.G.; Mier, M.Z.; Marks, J.C. (2009). Effects of travertine and flow on leaf retention in Fossil Creek, Arizona. *Hydrobiol.* 630: 187-197.

Copeland, R. (2003). Florida spring classification system and spring glossary. Florida Geological Survey Special Publication No. 52. Florida Geological Survey, Talahassee, FL.

Curry, R.A.; Brady, C.; Noakes, D.L.G.; Danzmann, R.G. (1997). Use of small streams by young brook trout spawned in a lake. *Trans. Am. Fisheries Soc.* 126 (1), 77-83.

Diéguez, C.; Hernández, J.M.; Pujalte, V. (2009). A fern-bennettitalean floral assemblage in Tithonian-Berriasian travertine deposits (Aguilar

Formation, Burgos-Palencia, N Spain) and its palaeoclimatic and vegetational implications. *J. Iber. Geol.* 35: 127-140.

Eckstein, Y.; Lewis, V.E.; Bonta, J.V. (2007). Chemical evolution of acid precipitation in the unsaturated zone of Pennsylvanian siltstones and shale of central Ohio. *Hydrogeol. J.* 15: 1489-1505.

Emig, W.H. (1917). Mosses as rock builders. *Oklahoma Acad. Sci.* 1: 38-40.

Ferretti, A.; Cavalazzi, B.; Barbieri, R.; Westall, F.; Foucher, F.; Todesco, R. (2012). From black-and-white to colour in the Silurian. *Palaeogeograph. Palaeoclimatol. Palaeoecol.* 367–368: 178–192.

Ford, T.D.; Pedley, H.M. (1996). A review of tufa and travertine deposits of the world. *Earth Sci. Rev.* 41:117-175.

Habdija. I,; Habdija, B.P.; Matoničkin, R.; Kučinić, M.; Radanovič, I.; Miliša, M.; Mihaljević, Z. (2004). Current velocity and food supply as factors affecting the composition of macroinvertebrates in bryophyte habitats in karst running water. *Biol. Bratislava* 59: 577-593.

Love, K.M.; Chafetz, H.S. (1988). Diagenesis of laminated travertine crusts, Arbuckle Mountains, Oklahoma. *J. Sediment. Res.* 58: 441-445.

McKee, J.K. (2010) Taphonomic processes of bone distribution and deposition in the tufa caves of Taung, South Africa. *J. Taphon.* 8: 203-213.

Mehran, M.; Arulanandan, K. (1977). Low frequency conductivity dispersion in clay-water-electrolyte systems. *Clays Clay Minerals.* 25: 39-48.

Munsell (2009) *Munsell Rock Color Book.* Munsell Color, Grand Rapids, MI. 14 pp.

Nagano. T.; Nakashima, S.; Nakayama, S.; Osada, K.; Senoo, M (1992). Color variations associated with rapid formation of goethite from proto-ferrihydrite at pH 13 and 40°C. *Clays Clay Minerals* 40: 600-607.

Pentecost, A. (2005) *Travertine.* Springer-Verlag, Berlin..

Potter, P.E.; Maynard, J.B.; Pryor, W.A. (1980). *Sedimentology of Shale: Study Guide and Reference Source.* Springer-Verlag, Berlin. 306 pp.

Ricker, W.E. (1931). Ecological study of the Mad River and notes on other Ontario streams with special reference to the speckled trout. University of Toronto, M.A. Thesis.

Riding, R. (2000). Microbial carbonates: the geological record of calcified bacterial-algal mats and biofilms. *Sedimentol.* 47: 179-214.

Rodie, A; Post, R (2009). Niagara Escarpment Baseflow Study. p. 59. Nottawasaga Valley Conservation Authority. Utopia, ON.

Rogerson, M.; Pedley, H.M.; Wadhawan, J.D.; Middleton, R. (2008). New insights into biological influence on the geochemistry of freshwater carbonate deposits. *Geochim. Cosmochim. Acta* 72: 4976-4987.

Scheinost, A.C.; Chavernas, A.; Barrón, V.; Torrent, J. (1998). Use and limitations of second-derivative diffuse reflectance spectroscopy in the visible to near-infrared range to identify and quantify Fe oxide minerals in soils. *Clays Clay Minerals* 46: 528-536.

Sewell, H.D. (1837). Travertine or calcareous tufa. *Trans. Lit. Hist. Soc. Quebec* 3: 255-260.

Steel, R.G.D.; Torrie, J.H. (1980). *Principles and Procedures of Statistics.* McGraw Hill, NY, p. 86-119.

Swatland, H.J. (1998). *Computer Operation for Microscope Photometry.* CRC Press, Boca Raton, FL. 238 pp.

Swatland, H.J. (2010). Polarized-light interferometry of calcium carbonate deposition in moss from a Niagara Escarpment waterfall. *Microsc. Microanal.* 16: 306-312.

Swatland, H.J. (2011). Microphotometry of underwater shadowing by a moss from a Niagara Escarpment waterfall. *Microsc. Microanal.* 17: 125-131.

Swatland, H.J. (2012). Groundwater temperature and degassing in the Mad River subwatershed of Lake Huron. *J. Great Lakes Res.* 38: 117-120.

Telford, P.G. (1976). Paleozoic geology of the Collingwood-Nottawasaga area. Ontario Division of Mines. Map 2341.

Tovell, W.M. (1992). *Guide to the geology of the Niagara Escarpment.* Niagara Escarpment Commission, Georgetown, ON. 159 pp.

Viles, H.A.; Naylor, L.A.; Carter, N.E.A.; Chaput, D. (2008). Biogeomorphological disturbance regimes: progress in linking ecological and geomorphological systems. *Earth Surf. Process Landforms* 33: 1419-1435.

Vos, M.A. (1975). Potential clay and shale resources of central Ontario. Ontario Division of Mines.

Webster, D.A., Eiriksdottir, G. (1976). Upwelling water as a factor influencing choice of spawning sites by brook trout (Salvelinus fontinalis). *Trans. Am. Fisheries Soc.* 105 (3), 416-421.

Weidner, V.R.; Hsia, J.J. (1981). Reflectance properties of pressed polytetrafluoroethylene powder. *J. Optic Soc. Am.* 71: 856-861.

Wilde, P.; Quinby-Hunt, M.S. (2010). Paradoxes and perceptions in color identification of paleo-redox conditions in peletic rocks from diagenic to metamorphic grade. *Int. J. Geol.* 2: 29-35.

Witzel, L.D.; Maccrimmon, H.R. (1983). Redd-site selection by brook trout and brown trout in Southwestern Ontario streams. *Trans. Am. Fisheries Soc.* 112 (6), 760-771.

Reviewed at NOVA SCIENCE PUBLISHERS, INC., 400 Oser Avenue, Suite 1600. Hauppauge, NY 11788-3619, USA

In: Groundwater
Editor: Segun Michael Adelana
ISBN: 978-1-63321-759-1

Chapter 9

GROUNDWATER DEPENDENT ECOSYSTEMS – SCIENCE, CHALLENGES, AND POLICY DIRECTIONS

David K. Kreamer*[*,1], *Lawrence E. Stevens*[2] and *Jeri D. Ledbetter*[2]
[1]Department of Geoscience, University of Nevada, Las Vegas, NV, US
[2]Springs Stewardship Institute, Museum of Northern Arizona, Flagstaff, AZ, US

ABSTRACT

Groundwater Dependent Ecosystems (GDEs) are foci of this planet's biological diversity, living records of cultural heritage, centers of religious ritual and practice, strategic goals of economic aspirations, and have been both the objects and weapons of local and international conflict. They have inspired poetry and the arts, and embody verdant byways in the most inhospitable deserts of the world and of the human spirit. Increasing exploitation of groundwater resources has interrupted the quantity and altered hydrochemistry of these springs and wetlands, creating resource management challenges around the globe. Basic information on GDEs is elusive, as many are in remote locations, and have small areal extent concealed from the resolution of both remote

[*] Corresponding author email: dave.kreamer@unlv.edu.

sensing satellites and supportive political action. Varying laws, policies, and management practices throughout the world have directed a path forward for GDE sustainability, amid conflicting national ambitions, constrained economic and enforcement capabilities, and lack of basic scientific data.

Keywords: Groundwater Dependent Ecosystems, Springs, Policy, Cultural Significance

INTRODUCTION

In a "Report to Congress - Progress Toward Establishing a National Assessment of Water Availability and Use" (Alley et al. 2013), the work conducted on the U.S. National Water Census and future directions of the program are given a clear and wide-ranging overview. What is remarkable about the document is the limited discussion of groundwater in general, and the relative absence of any discussion of the relationships of ecosystems and groundwater in the vast arid and semi-arid regions of the United States. Criticisms of the Report's drafts, which pointed out there was little or no mention of springs, seeps, groundwater dependent wetlands or river reaches, and subterranean ecosystems, were deemed by the authors to be not in keeping with the overall theme of the report and the technical level to which the report was to be written. That lack of mention of the dependence of ecological systems on groundwater is echoed in the dearth of laws, policy initiatives, scientific study, and public attention on the issue, not only in the U.S., but in many countries throughout the world. Some countries, like Zimbabwe, have robust and holistic water laws, but have difficulties in their implementation and management (Makurira and Magumo 2003). With burgeoning worldwide population, more and more pressure has been put on groundwater for irrigated agriculture, industry, and municipal supply with resultant overdrafts and diminishment of environmental flows and chemistries. For example, over half of India's agriculture is irrigated with groundwater and that fraction is increasing (Dubash 2002), putting pressure on groundwater dependent springs. Immense portions of the planet's population subsist on the limited and vulnerable water that leaks to the surface, or is withdrawn from wells and boreholes. Particularly in dry regions of the earth, the productivity, biological diversity and cultural importance of groundwater dependent ecosystems (GDEs) are often orders of magnitude greater than surrounding dry areas that are not directly supported by groundwater (Odum 1957, Stevens and Meretsky 2008).

Defining and sometimes constraining public perception of these ecosystems are the varied and individual laws of many countries, regions and states. These laws and regulations direct the public policies and management approaches to either recognize or to ignore the relation of groundwater to life-forms it sustains. In turn, scientific research, characterization and monitoring of GDEs relies on funding and supportive public policy. Where clear and sustaining laws, regulations, policies, and monitoring are not available, important ecosystems can suffer and biological diversity can be lost. The efficacy of enforcement of those laws and policies toward natural resource conservation and groundwater dependent ecosystems is key in protection and sustenance of many unique, and sometimes fragile, ecological communities.

Compounding these problems, the locations of springs, seeps, and other GDEs are often not well characterized, even in economically advantaged societies that can apply resources for scientific study. Management of, and policy development for, GDEs cannot go forward if springs, seeps, wetlands, gaining reaches of streams, groundwater-dependent terrestrial ecosystems and phreatophytes, aquifer ecosystems, and cave ecosystems have not been located, identified, surveyed, and monitored. Unfortunately, the basic value of GDEs, and even the existence and enumeration of these unique biotic communities has not been realized in many regions, and the paucity of information constrains public perception that, in turn, drives proper management and policy development. Finally, the definitions of the variety of GDEs are inconsistent and not evenly applied either in law or in practice (Springer and Stevens 2008).

THE SIGNIFICANCE OF GROUNDWATER DEPENDENT SPRINGS AND WETLANDS IN HUMAN HISTORY

GDEs and the groundwater that supplies them have been pivotal in human history - our culture, religions, and traditions are intimately linked with these places. Many past and present civilizations have located, and had agriculture flourish, near wetlands including: Egyptian society and the periodically flooding Nile Valley, the Khmer Empire and the Mekong Valley particularly around Tonle Sap (Great Lake) and the ancient hospital of Neak Pean in Cambodia, Rome's ancient port of Ostia Antica, Spain's Empúrias in

Catalonia, the ancient western Greek city of Nicopolis in the Amvrakikos Gulf, and Mesopotamian civilization in the Tigris and Euphrates lowlands (Ramsar 2008). Major cities such as Venice, Bangkok, Mexico City, Tunis, Guayaquil, Amsterdam, New York, and New Orleans were established near existing wetlands, although the threat of malaria in some parts of the world have driven human population away from some wetlands (Ramsar 2008). Archaeologically, the anoxic and waterlogged conditions associated with peatlands have yielded valuable historical finds. Ancient ships have been exhumed from Venice and Marseille, and Etruscan age vessels have been recovered from the Hutovo Blato wetland in Bosnia and Herzegovina. Ancient and pre-historic settlements have been discovered in peatlands of the United Kingdom, in Black Sea bottom muds, and under Lake Titicaca. In addition to preserving human relics, the anoxic conditions of peatlands can preserve past plants, insects and pollens, which can serve as valuable indicators of past climate (RAMSAR 2008).

Additionally, springs and wetlands have served as focal points of hostilities, national ambition and war, as both an object of conflict and a strategic tool. For example, Gihon Spring is one of several springs that served as the original gathering locations for what was to become the city of Jerusalem. These springs, located just outside the walls of the early, growing city, were not only major points of sustenance, but were manipulated for strategic military purposes. As mentioned in the King James Bible, when Assyrians were advancing toward Jerusalem in 701 BC, the water flow for these springs was purposely stopped up to impede them (Scofield 1967).

Religiously motivated wars have involved springs and groundwater as strategic tool. In one example during the Crusades in mid-summer of 1187 AD, 30,000 Muslim forces, including 12,000 cavalry, were approached by an army of 20,000 Crusaders at the Horns of Hattin (near Tiberias in present day Israel) having been lured away from the springs at Saffuriya. The Crusaders last sure source of water was a spring at the city of Tur'an, 14 km from their objective of Tiberias. When the Crusaders set out at mid-day, the Muslim leader Saladin ordered two wings of his army to take the spring at Tur'an and cut off the retreat and water source of the advancing army. The then thirsty Crusaders veered toward the springs of Hattin, and they were met and stopped in fierce battle. Saladin's armies captured or killed the large majority of the Crusaders who had reduced the garrisons of their castles and fortified cities to field their army, making Islamic forces the foremost military power in the Holy Land and prompting a Third Crusade (Gibb 1969, Madden 2000).

Locations of GDEs are often considered as sacred sites, and are crucially important to cultural and religious practice. In ancient Greece, visitors typically stopped at the Castalian Spring near Delphi, which rests in a narrow valley between the "shining" cliffs called Phaedriades, near the Temple of Apollo. Roman poets traveled great distances to receive inspiration at the spring, as did the contestants in the precursors to the modern Olympic Games, the Pythian Games. The spring was also used by supplicants visiting the nearby Delphic Oracle, a cave springs system considered by Hellenes to be the center of the earth, where legend indicates that Apollo slew the chthonic earth guardian monster Python, and a source of divine knowledge and guidance (Broad 2007). Other examples of springs' association with the sacred include the location of monuments to the Babylonian moon goddess Ishtar; the Gallo-Roman Fontes Sequanae at the source of the Seine River in France; the Christian Grotto of Lourdes at the foot of the Pyrénnées in France; the mythic "Fountain of Youth" sought out by the Spaniard Ponce de Léon in Florida, USA; the 7000 year old mineral springs in Bath, England with the Celtic shrine to Sulis and the Roman shrine to Sulis Minerva; and the life-giving fountain and spring waters of Pon Lai and Mount Lao Shan in China (Witcombe 1998).

On the other side of the globe, ancient and modern day Native Americans regard water issuing from the ground as central to their religious beliefs and practices. The hydrologic cycle of indigenous Native Americans is deeply invested with spiritual life. Iconographic analyses across the southern United States and northern Mexico document pictographs of a horned serpent that is responsible for bringing water to the surface of the earth and guarding springs (Phillips et al. 2006, Rea 2008). Also like Delphi, several springs in the southwestern U.S. are considered to be sites where humanity emerged into the world, while at least one Tribe considers a springs complex as heaven, where the spirits of the dead dwell in perpetuity (Hart 1986). The Mayan civilization, in what is now Mexico, viewed caves and cenotes as "doors to the underworld" or U'kux Xibalba. Virtually all indigenous cultures on Earth revere groundwater dependent surface or subterranean flows as sacred places where miracles occur.

Springs have been intensively used by indigenous cultures for >10,000 yr in the U.S, and far longer in Europe, Africa, Asia, and Australia (Haynes 2008). Consequently, springs also emerge as important paleontological and archeological sites for understanding landform, human, and cultural evolution. The affiliation of humans with springs over evolutionary time requires that those involved in spring ecosystem restoration carefully consider the roles of both natural and anthropogenic disturbance (Kodrick-Brown and Brown 2007).

THE IMPORTANCE AND CHARACTER OF GROUNDWATER DEPENDENT ECOSYSTEMS

Recognizing that informed policy development and proper management of GDEs can not go forward if the value and existence of these vital resources is undefined, it is crucial that planners and decision-makers understand their importance. Springs are emblematic of the rich diversity and importance of GEDs. Springs are often point-sources of biodiversity within landscapes, providing essential habitat for many endangered species, and supporting untold hundreds of rare or unique plant, macroinvertebrate, fish, amphibian, and some reptile and mammal species. For example, more than 200 species of hydrobioid spring-snails in the genera *Pyrgulopsis, Trionia,* and lithoglyphid snails in the genus *Fluminicola* occur in springs and spring-fed stream channels throughout North America, and at least 23 species of in the genera *Jardinella, Caldicochlea, Fonscochlea, Trochidrobia*, and *Austropyrgus* are endemic to the Great Artesian Basin of Australia (Hershler 1994; Hershler and Frest 1996; Liu et al. 2003; Perez et al. 2005; Hershler and Liu 2008, 2012; Murphey et al. 2012). Other aquatic gastropod families, such as Physidae also contain large numbers of springs-dependent species (Wethington and Lydeard 2007). Numerous springs-endemic fish also exist in arid portions of the United States and Australia, including the families of Cyprinidae, Cyprinodontidae, Goodeidae, Poeciliidae, Gobiidae, and other families (Kodrick-Brown and Brown 1993, 2007; Echelle, et al. 2005).

Individual springs may support high levels of unique species (e.g., Montezuma Well in central Arizona, USA) (Blinn 2008). Spring-dominated valleys, such as Cuatro Cienegas in northern Mexico and Ash Meadows in southern Nevada, USA, support dozens of unique plant, invertebrate, fish, and other vertebrate species (Hendrickson et al. 2008; Williams 1984). However, there is often little recognition of the extent of endemic species' dependence on springs. For example, nearly all of the 12 endemic plant species in Ash Meadows, Nevada exist only in groundwater-fed alkaline meadows. In addition, springs sometimes provide important habitat for critical phases of life history, such as overwintering in warm water by Florida manatees, *Trichechus manatus latirostris* (O'Shea and Ludlow 1992), hibernacula of eastern bog turtles, *Glyptemys muhlenbergii* (Carter et al. 1999) in the northeastern U.S. or the in-snail phase of the trematode flatworm *Leucochloridium cyanocittae*, which occurs in Arizona only in *Oxyloma haydeni* ambersnails in springfed wetland vegetation at Vasey's Paradise in Grand Canyon National Park, Arizona, USA (Meretsky and Stevens 2000). The role of spring waters in imprinting among larval fish also may be substantial, but has not been studied to our knowledge.

Some springs are highly productive, and despite their small size, often contain tightly packed assemblages of plants and animals. In a recent study of springs in Kaibab National Forest in northern Arizona, USA, we detected 20% of the entire forest flora in <0.001% of the landscape, representing a remarkable concentration of plant species richness at springs (Ledbetter et al. in press). We found similar results in the Spring Mountains, Nevada, USA (U.S. Forest Service 2012c). Similar species packing has been noted among mosses, aquatic beetles, and avifauna (Grand Canyon Wildlands Council, Inc. 2002), and is a characteristic of subaqueous springs as well (McClain et al. 2003). Thus, everywhere, the high productivity and ecological stability of spring habitats likely contributes to tighter species packing.

Springs that exhibit high concentrations of endemic species are likely ancient systems, and may be trophically strongly structured (Kodric-Brown and Brown 1993; Blinn 2008). The evolution of endemism occurs regularly at springs, arising through relictualization and adaptation cycles related to environmental change: Onset of an altered environmental regime (e.g., desertification, marine or freshwater transgression, or other large-scale environmental changes) may reduce distribution and availability of habitat, enforcing isolation, allele fixation, and local adaptation. In contrast, habitat expansion permits widespread distribution of formerly isolated populations into new habitats. For example, Neogene lakes in the Great Basin of the American Southwest underwent extensive expansions and contractions under changing glaciation-related climates since the Miocene, contributing to the adaptive radiation and now highly localized endemism of spring-snails, pupfish, and aquatic Hemiptera (Hershler and Liu 2008; Echell et al. 2005, Polhemus and Polhemus 2002). Non-native species studies of weedy plants, Gastropoda, crayfish, amphibians, and fish (Stevens and Ayers 2002; Sada et al. 2006). reveal that spring ecosystems are particularly prone to invasion, a condition that has had evolutionary consequences for numerous lineages.

In addition to their role in regional biodiversity and evolution, springs often function as keystone ecosystems playing disproportionally large roles in adjacent upland ecosystems (Perla and Stevens 2008). Aridland and mesic land springs alike provide essential water, food, and habitat resources for large numbers of invertebrates, fish, amphibians, reptiles, birds, and some mammals. Landsnails, ground beetles, butterflies, and both migrant and nesting birds all are found in great concentrations at springs in the American Southwest. Springs and wetlands also may serve as refugia during drought for terrestrial animal species, and as regeneration hotspots following fire in forested regions (Grand Canyon Wildlands Council, Inc. 2002). Wetlands can clean and filter surface water, serve as sinks for greenhouse gases, and, particularly in estuarine systems, sequester carbon from the atmosphere (Whiting and Chanton 2001; Bridgham et al. 2006). In contrast, some springfed playas may serve as CO2 sources, as alkali soils may oxidize organic matter (Crews and Stevens 2009). Overall, GDEs in general and springs specifically encompass intricate and typically biodiverse ecosystems, and can be found in subaqueous and subaerial settings in marine, coastal, lotic, lentic, wetland, terrestrial, cave, and aquifer environments.

Science and GDE Ecosystem Classification

Prior to the 21st Century, the classification and ecology of groundwater dependent ecosystems was largely ignored. However, that information gap is rapidly being filled. In Australia, much attention has recently been devoted to the classification of near-surface GDEs, as summarized in Hatton and Evans (1998) and Eamus et al. (2006). These approaches emphasize the proximity, availability, flux, and quality of groundwater to the surface. The U.S. Forest Service also recently refined classification to improve forest springs inventory of GDEs (U.S. Forest Service 2012a). However, GDE classification is problematic, in part because the phrase encompasses ecosystems deep in the crust to nearly all surface aquatic and some ice-bound environments on Earth: shallow and deep aquifers, springs, streams, rivers, lakes, and oceans may all qualify as GDEs. Among this array of ecosystems, springs are one of the most obvious groups of GDE systems; however, classification of springs has been largely ignored by the public, the scientific community, and most government agencies. Neglect of springs is demonstrated by the near absence of reference to springs in most contemporary texts and textbooks on national or global freshwater ecosystem health, wetland ecosystems, and water resource management. Our review of existing literature suggests several reasons why springs have received so little attention, including issues of definition, scale, and basic understanding.

1) Meinzer (1927) provided guidance on springs hydrology but not ecosystem ecology, and until recently, the inadequacy of an ecosystem-based classification system has precluded discussion of springs types and thwarted mapping. Springer and Stevens (2008) described 12 types of terrestrial springs, of which several also appear to be common subaqueous forms. Spring types vary in abundance in different geologic provinces; however, the absence of a classification system has largely prevented discussion of variation among spring types across spatial scale. Fens in particular have posed conceptual problems for springs classification. Although all fens may qualify as wetlands, not all are groundwater-fed. Those that are groundwater-fed, but exist at low elevations are often called *ciénegas* or *vegas* , while elsewhere they may be called wet meadows, fens, or helocrenes. Classification of springs is important so that rare types can be identified and studied.
2) Springs are commonly overlooked in landscape resource mapping projects because they are usually small and fall within, rather than among, mapping pixels. While highly variable, the average area of individual springs in western North America is 1861 (n = 526; 1 sd = 18962) m^2, with few springs habitat patches larger than 1 ha (Ledbetter et al. 2012). Also, while the mean discharge of western North American springs was11.8 L/s, (n=510; sd=151), the discharge of 62% of springs is in the range of 0.01-1.0 L/s, with a median of 0.07 L/s. Thus, springs are small in area and discharge, and fall below the scale of most landscape mapping efforts and are neglected in remote sensing and floristic mapping.
3) Springs ecosystems have been overlooked by the science community, in part because of their ecological complexity. We note that the often high biological diversity in springs ecosystems is due, in part, to the co-occurrence of up to 13 different microhabitats within some complex springs types, such as Indian Gardens and large gushet springs, like Thunder River, both in Grand Canyon in Arizona. Groundwater-dependence, surface-linkage, microclimate complexity, intricate mosaics of co-occurring microhabitats and associated assemblages, and complex land use histories are characteristics that require detailed interdisciplinary studies. However, such studies are difficult to organize and fund, and with a few exceptions remain largely outstanding. Consequently, there is relatively little literature that integrates information on springs' ecosystem ecology, and the

best known text books on wetlands, ecosystem ecology, and the state of the world's ecosystems largely ignore springs. Further, the impacts of drought, water table decline, flow diminishment, and water quality change do not impact all phreatic or wildlife species in the same way, complicating resource management (Eamus and Froend 2006).

In spite of their ecological and economic importance, particularly in arid climates, mapping of springs is insufficient. Throughout North America and Europe, tens of thousands of farms, small communities, and many towns and some cities obtain their potable water from springs (Kresic and Stevanovic 2010); however, many of those springs have not been mapped. Only a few states within the U.S. have adequate maps of their springs, despite high densities of springs in montane regions (Stevens and Meretsky 2008). Arizona, Florida, Kentucky, Missouri, Nevada, New York, Utah, and Wisconsin are among the states with moderately good springs maps. Arizona, the second-driest state in the U.S. likely contains the highest concentration of springs, although the total number is still under investigation (Ledbetter et al. 2012). Several geographical information system (GIS) layers exist for Arizona, but none are complete, and each contains features not included in the others. For example, 141 springs are referenced in various databases available for Grand Canyon National Park (National Hydrography Dataset, Geonames, Arizona State Land Office, Arizona Department of Water Resources, U.S Geological Survey maps). However, this is no more than 23% of the known springs there, the others having been reported by independent researchers and in other publications. Unfortunately, such data gaps are the norm, rather than the exception, and springs remain poorly mapped worldwide.

Remote sensing has been promoted as a possible means of accelerating springs mapping. Some springs types are relatively easily detected using remote sensing methods (i.e., helocrene, limnocrene, and some low-gradient hillslope springs in non-forested terrain), while springs types associated with cliffs, caves, and other high-relief landforms (e.g., hanging gardens, gushets) and those emerging in heavily forested landscapes are difficult or impossible to detect using remote techniques. Recognizing these challenges, GDEs were located in the Western Cape of South Africa, using LandSat TM imagery and then mapping and modeling those sites with GIS approaches (Munch and Conrad 2007). Great Artesian Basin springs in Australia were mapped using remote sensing technology with good success (Lewis et al. 2013); however, similar efforts in other regions have met with mixed results. Detection accuracy for western U.S. springs ranged from 50.7% in areas with steep terrain and heavy forest cover (e.g., in the Spring Mountains of southern Nevada) to 83.8% in areas with low topographic relief and large fen wetlands (e.g., in Rocky Mountain meadows; Remote Sensing Applications Center 2012). In such landscapes, rigorous compilation of existing information and structured searching will likely be required to accurately map springs. Much basic mapping still needs to be conducted, both in terrestrial and subaqueous settings.

If surface GDEs are poorly recognized and understood by the scientific community, the public, and policy makers, the importance of subterranean groundwater ecosystems is far less well understood. Humpheys (2006) points out the great variety of life in aquifers, particular karst environments, including "diverse metazoan faunas comprising obligate groundwater inhabitants, largely crustaceans but also including insects, worms, gastropods, mites and fish". Sometimes these are endemic organisms of limited range, are vicariant from closely related biotic strains because of geological sequestration, and may be relictual in lineage, which allows examination of the history of deep aquifers (Humphreys 2006). In arid and/or coastal regions, saline waters can host obligate groundwater fauna (stygobites). Importantly, these ecosystems can be very vulnerable to changes in groundwater flow, quality, or habitat changes.

Clearly, the close relationship of groundwater flow/quality to both terrestrial and aquatic ecosystems is poorly understood by the public, much of the scientific community, and most government agencies. Several cultural factors contribute to this lack of awareness. Historically, the development of nations typically began near regions of available water, often near rivers or in more humid regions (where possible) that were less dependent on groundwater (although many rivers have gaining reaches that are effectively supplied by groundwater). Thus legal development, social awareness, policy actions, and resource management did not typically focus on groundwater in the early development of most nations. This cultural "surface water" focus of early population development is particularly evident in western Africa and South America, where many nations have the first major cities formed in wetter, coastal regions or along rivers, rather than the inland, more remote or drier regions. As populations increased in drier regions, groundwater usage, pollution, and overdraft became more common. Commensurate with arid-lands population growth, the legal, policy-making, and resource management protocols for groundwater and its dependent ecosystems had to evolve after-the-fact, from seats of power often far removed. Even though the population and climate change stresses on groundwater are more evident each year, laws, policies, funding for scientific research, and management strategies for preserving the quality and quantity of environmental flows lag behind and fail to accommodate human demands for drinking and irrigation water.

LAWS, POLICIES AND ENFORCEMENT – A GLOBAL PERSPECTIVE

Crafting and interpreting of water-related laws for the protection of GDEs takes many different forms throughout the world. The overriding legal structure and strictures of each country, and how existing laws are enforced and interpreted, sets the tone for policy decisions and gives a regional context for safeguarding springs, seeps, and wetlands. But many early laws and regulations addressed, at the most, a limited range of groundwater or ecosystem scenarios. The first international legal instrument to address transboundary groundwater was the Helsinki Rules of 1966, issued by the International Law Association. These Rules closely linked groundwater to surface water drainage systems and watersheds, but did not include all groundwater (Salman 2009). According to Lazarus (1997), because of the clear connection of groundwater and surface water, their bifurcation in common-law is, "clearly artificial and at odds with hydrological reality." He points out that most modern jurisdictions, at the time of his writing, drew no distinction between surface and subterranean waters.

No matter what the legal format, interpretation of law is crucially important for the preservation of ecosystems. In his seminal 1972 book, "Should Trees Have Standing? – Law, Morality and the Environment " Christopher Stone argues that affording legal rights or "standing" to natural objects in the world would support their protection (Stone 1972). "Standing" or *locus standi* is defined as "the ability of a party to bring a lawsuit in court based upon their stake in the outcome," and, " A party seeking to demonstrate standing must be able to show the court sufficient connection to and harm from the law or action challenged" (USLegal 2014). By affording environmental entities such as trees (or more broadly in our case Groundwater Dependent Ecosystems) "standing", Stone means that citizens could bring court action on behalf of harmed or threatened natural objects, and he points out that the courts already typically allow guardianship provisions, which permit third-party representation of those with standing who are unable to represent themselves. Countries have not adopted this interpretive legal strategy; however, some seek, through their legal systems to preserve the natural world to varying degrees.

Several nations have devised laws that have direct philosophical links to protection of unique ecosystems, including those habitats sustained by groundwater, but there are many local and regional differences. In particular, Africa is a continent with a variety of water and ecosystem laws, a multiplicity of agency and political support and/or interference, and a range of abilities to enforce existing laws. For example, before the 1990s, South African water law consisted largely of piecemeal allocation of water rights to individuals and private entities, but with the end of Apartheid, fundamental approaches to resources were altered. The South African Water Act of 1998, in addition to establishing availability of clean drinking water as a basic human right, establishes water as an integral part of ecosystem protection. Among the goals of the Act are clear environmental priorities, in addition to general objectives of equitably meeting basic human needs and redressing the results of past racial and gender discrimination. The South African Water Act of 1998 specifies the goal of "promoting the efficient, sustainable and beneficial use of water in the public interest" and even more pertinent to this discussion, "protecting aquatic and associated ecosystems and their biological diversity". It is argued by some authors that this historic water law took some of its early guidance from preceding improvements in Australian water law, but now a few years after its enactment, the roles might be reversed and "Australia can now look to the new South African legal and constitutional framework for inspiration for reforms more geared towards sustainable environmental and social outcomes" (Godden, 2005). The Act allows any citizen to be an environmental advocate.

Other African States have refined their legal framework for water law as well with varying degrees of success for explicit or implicit protection of GDEs. In 1999 Kenya created a "National Policy on Water Resources Management and Development", which was followed by the Water Act of 2002. These initiatives set a tone for integrated water management (Mutinga, 2005), which did not exclude the protection of GDEs and allowed for discussion of the need in Kenya for biodiversity (Khamala 2005). Likewise in 1998, after extended dialogue with stakeholders, Zimbabwe established a new Water Act. Interestingly, both surface water and groundwater are treated as part of one hydrological system in the Act and there are provisions for environmental sustainability. Makurira and Magumo (2003) observed that in the first five years, Zimbabwe's water law had not reached its full potential for several reasons, including financial instabilities, the withdrawal of several donors who initially supported the program, other interfering national initiatives, inadequate agency and institutional linkages and low resources, lack of enforcement of the law, and political obstruction. Botswana's Water Law of 1968 defines underground water as public water, without property right (Lazarus 1997). In Ghana, a Water Resources Commission (WRC) was created by an Act of Parliament (Act 522 of 1996) in order to integrate water resources management and "regulate and manage the country's water resources and coordinate government policies in relation to them" (Odame-Ababio 2003). The Commission's recent groundwater strategy for Ghana has a strong emphasis on resource conservation sustainability and supply for the population, but has little direct emphasis on groundwater protection for ecosystem safety. The Canadian International Development Agency has been providing financial support for the Hydrologeological Assessment Project in the northern regions of Ghana in assistance of WRC efforts. While the Ghanaian WRC is rigorously evolving toward a decentralized stakeholder engagement model for groundwater management, the preservation of GDEs are currently not explicitly stated, nor are stakeholders explicitly assigned to consider GDE sustainability (Ghanaian WRC 2011).

Legal protection of springs and other GDEs takes different forms in other countries. In the United States, the Endangered Species Act of 1973 (ESA) is a major force that has been exercised many times to protect ecosystems and threatened species (Nelson 2008). In interpreting the law and responding to legal challenges, the U.S. Supreme Court ruled, in a landmark and precedent-setting case, that "the plain intent of Congress in enacting" the ESA "was to halt and reverse the trend toward species extinction, whatever the cost" (Tennessee Valley Authority vs. Hill 1978). The U.S. Fish and Wildlife Service (FWS) and the National Oceanic and Atmospheric Administration (NOAA) administer the ESA. Australia developed its own Endangered Species Protection Act in 1992, which was subsequently strengthened with replacement by the Environmental Protection and Biodiversity Conservation Act of 1999 (enacted in July of 2000). Australia also has instituted a National Strategy for Ecologically Sustainable Development, a National Water Quality Management Strategy, and the Water Act of 2007, which includes provision for groundwater and environmental flows. In many ways, the broader view of ecosystem protection, afforded in Australian law, can allow better overall management than the legal drivers in the U.S. that directs a more piecemeal, endangered-species by endangered-species, non-integrative approach.

Various, and sometimes different legal approaches and definitions have historically been applied to both surface waters and groundwaters that support

GDEs. Many nations place water in the public trust, at either provincial/state or federal levels. Provincial or state regulation of interstate aquifers may require complicated political negotiations that may fail to achieve sustainable management of those aquifers. However, there are numerous examples of nations with an integrated, holistic approach that treat both above and below ground waters as part of the same hydrologic cycle. For example, the Israeli Water Law of 1959 confers all water, both above and underground to the State; likewise, the Spanish Water Law of 1985 defines groundwater resources as public property (Lazarus 1997). Broad-reaching legal definitions of water are contained in the New Zealand Resource Management Act of 1991 that protects water in all forms, both above and below ground, as does the Alberta (Canada) Water Resources Act of 1979, amended in 1981. The British Colombia Water Act of 1979 (Canada) a “stream” is characterized as “a natural watercourse or source of water supply, groundwater, a lake, river, creek, spring, ravine, swamp and gulch” (Lazarus 1997).

The European Union actively recognizes that there is not enough information on GDEs, and has created policies and funding to advance a scientific groundwater research effort under the European Communities Framework Programmes for Research and Technological Development. Specifically the Framework 7 (FP7) directive includes a sub-program with an acronym of GENESIS, funded at 7 million Euro, which supports and encourages research on groundwater and GDEs in the European Research Area (ERA). The project timeline is April of 2009 to the end of March 2014, and includes participating institutions from 17 countries (CORDIS 2014). A particular appreciation of the EU effort, that groundwater dependent ecosystems need to be better understood in Europe, is advancing both scientific understanding and public awareness of these vulnerable spring and wetlands. Even in laws and legal constructs preceding these more recent actions, European nations recognized surface and groundwater as interconnected, with an example being the German Federal Water Amendment Act of1986 (Lazarus 1997).

There have been many international initiatives addressing water related ecosystems in the last half century. Three of the most influential are the Ramsar Convention of 1971, the already mentioned European Union’s Water Frame Work Directive (WFD; EU Directive 2000/60/EC), and the Brisbane Declaration of 2007. The former was written and adopted by participating countries at a summit in Ramsar, Mazandaran, Iran and takes an expansive approach to protect and influence the future of wetlands. The broad definition of wetlands in Article 1.1 of the Convention is “areas of marsh, fen, peatland or water, whether natural or artificial, permanent or temporary, with water that is static or flowing, fresh, brackish or salt, including areas of marine water the depth of which at low tide does not exceed six metres" (Ramsar 2014). This broad definition focuses on surface water, including even salt pans, fish ponds and rice paddies, but also recognizes the importance of groundwater. A Ramsar list of "Wetlands of International Importance" now numbers 2,171 sites, covering 207,291,271 hectares with many in the United Kingdom and Canada. According to the Ramsar website (Ramsar 2014) there are 168 contracting parties, as opposed to only 21 original signatory nations in 1971. The European Union's Water Frame Work Directive has perspectives that fundamentally redefine the approach to bodies of water, both above and below ground. The Directive promotes attainment of "good statPa Der Vang <pdvang@stkate.edu>

us" through an assessment aimed at a water body's ecology and chemistry. An important part of the determination of groundwater status is ascertaining the potential impairment of ground-water dependent terrestrial ecosystems (Krause 2007). A strength of the directive is its holistic and inclusive approach, its main weakness likely that it is organized around surface river basins which do not always conform to aquifer boundaries. The more recent Brisbane Declaration was released at the 10th International River Symposium and Environmental Flows Conference, held in Brisbane, Australia, in September 2007. Although the Declaration is a powerful and urgent call to the global community for the defense of imperiled freshwater and estuarine ecosystems, there is scant mention of groundwater dependent habitat. The Declaration does, however, emphasize "environmental flows", both in quantity and quality, as a factor crucial to the future of GDEs.

Hydrologic Prediction, Ecological Evaluations, and the Future of Groundwater Dependent Ecosystems

The solutions to GDE protection lie in increased scientific scrutiny, shifts in the tectonic plates of public perception, and strengthened policy decisions. Recent scientific advances and better monitoring of GDEs, speed the process. Traditionally, a comprehensive assessment of the chemistry and physical properties of groundwater issuing to an ecosystem is a necessary first step. Along with traditional groundwater tracking, tracing and dating techniques, new methods are actively being proposed to establish flow source and potential vulnerabilities to host springs (Kreamer et al. 1996; Johannessonn et al. 1997; Ingraham et al. 2001; Heilweil et al. 2014). These techniques use water quality parameters to track and trace groundwater status, movement, sustainability, and age. The proper amount and chemistry for individual GDEs varies, and several researchers such as Eamus and colleagues (Eamus et al. 2006a; Eamus et al. 2006b) have proposed methodologies for determining the groundwater flow and quality needed to successfully supply an ecosystem.

The development of organized and methodical GDE assessment techniques have been a major managerial advance in the last decade. Recent refinements in springs classification and monitoring allow a more systematized approach to their evaluation with Springer and Stevens (2008) developing 12 different spring types based on their hydrogeological setting and spheres of discharge. Boulton and Hancock (2006) have looked at groundwater-dependent rivers, determining both the degree of dependency, and the influence of upwelling hyporheic waters on the physical and chemical riverine processes. They discuss the important management implications of understanding and predicting the baseflow contributions to river biota and its quantitative sustainability. Integrative groundwater ecosystem assessment is put forward by Stuebe et al. (2009) and the Springs Stewardship Institute (SSI 2014) pointing out the need of determining natural background, establishing bioindicators (perhaps microbial populations) for understanding ecosystem health, and attempting to gain correlation between abiotic and biotic variables.

One recent, exciting example of an organized GDE inventory protocol comes from the U.S. Forest Service. In an organized effort to promote the best available information and science, three intensity levels of GDE inventory and monitoring were established (USFS 2012a; USFS 2012b). The first level is geared toward referencing the location and extent of GDEs. The second involves the description of major attributes of GDEs such as flora and fauna, the hydrogeologic setting, factors which could perturb the site. The third level centers on gathering extremely quantitative spatial and temporal information to characterize physiochemical aspects of a GDE. The approach includes considering business required analysis and protocols (USFS 2010). These guides address many considerations including: 1. pre-field survey activities such as mapping, remote sensing, and gathering geographic information, 2. field survey activities such as use of a Management Indicator Tool, georeferencing, transect layout and describing the geology, soil, hydrology, vegetation aquatic and terrestrial fauna, and any natural and anthropogenic disturbance, and 3. post-field activities such as laboratory analyses, information verification and data interpretation. These protocols have been used and documented at many sites.

Another new management tool is the classification of ecosystem resilience and/or vulnerability. For example, originally, simple and subjective desktop methods for GDE assessment were available such as a 5- or 10-point scale ranking of uniqueness, vulnerability and priority (Nature Conservation Council 1999; Sinclair Knight Mertz 2001; Murray et al. 2003). Since that time, more comprehensive and complicated modeling approaches have been used. An evaluation of ecosystem resilience was made in regions transected by the Qinghai-Tibet railway in China using GIS, and using a neural network to analyze ecosystem risk (Chen et al. 2007). In their studies of wetland vulnerability in the U.K. and Ireland, Krause et al. (2007) put forward an ecohydrogeological framework for assessing groundwater dependent terrestrial ecosystems, including both their chemical and ecological status. This is accomplished by distinguishing wetlands based on controlling ecological, hydrological, chemical and hydrogeological characteristics, and then assessing the potential and risk of damage. The SSI (2014) springs ecosystem assessment protocol advances these concepts and approaches further, by using quantitative data and expert opinion to provide springs stewards with within- and among-site management priorities.

Perhaps more daunting is the challenge of changing community perceptions. GDEs can be negatively viewed as breeding grounds for mosquitoes and other pests, unutilized land that should be drained, swampy areas that attract detrimental and nuisance wildlife and create odors that degrade the value of surrounding land, and/or resources that should be piped to another place. Many communities do not recognize the connection between groundwater overdraft, and surface water diminishment, while others can not appreciate the link between groundwater and surface water quality.

The push to increase public appreciation and understanding of GDEs is moving forward, and many groups are joining in this effort. For example, the International Association of Hydrogeologists has established a "Groundwater and Ecosystems Network" to raise public awareness and promote the welfare of GDEs, hoping to positively influence local, national and international policy (IAH 2014). Protection of wetlands through the efforts of the Ramsar Convention are increasingly broad and effective, and the Springs Stewardship Institute (2014) is working to advance interaction and training among springs managers to improve understanding and management of these resources (Ramsar 2014; Spring Stewardship Institute 2014). Many wildlife societies are joining in the effort to protect groundwater, which issues into areas of unique biological diversity and habitat. The success of these and similar activities will be key in preservation of these priceless oases, scattered around the globe, that provide such rich and varied habitats for groundwater dependent species, human cultural resources, and essential ecosystem functions.

REFERENCES

Alley, W.M., Evenson, E.J., Barber, N.L., Bruce, B.W., Dennehy, K.F., Freeman, M.C., Freeman, W.O., Fischer, J.M., Hughes, W.B., Kennen, J.G., Kiang, J.E., Maloney, K.O., Musgrove, M., Ralston, B., Tessler, S., and Verdin J.P. (2013). Progress toward establishing a national assessment of water availability and use: U.S. *Geological Survey Circular* 1384, 34 p., available at http://pubs.usgs.gov/circ/1384.

Blinn, D.W. (2008). The extreme environment, trophic structure, and ecosystem dynamics of a large, fishless desert spring: Montezuma Well, Arizona. Pp. 98-126 *in* Stevens, L.E. and V.J. Meretsky, editors. Aridland Springs in North America: Ecology and Conservation. University of Arizona Press, Tucson.

Boulton, A.J., Hancock, P.J. (2006). Rivers as groundwater-dependent ecosystems: a review of degrees of dependency, riverine processes and management implications. *Australian Journal of Botany* 54(2) 133–144 http://dx.doi.org/10.1071/BT05074

Bridghan, S.D., Megonigal J.P., Keller J.K., Bliss, N.B., Trettin, C. (2006). The Carbon Balance of North American Wetlands. Wetlands Vol. 26, No. 4. Pp. 889-916.

Broad, W.J. (2007). The Oracle: Ancient Delphi and the Science Behind Its Lost Secrets. Penguin Press, New York.

Carter, S.L., C.A. Haas, Mitchell J.C. (1999). Home range and habitat selection of bog turtles in southwestern Virginia. *Journal of Wildlife Management* 63:853-860.

Chen, H., Liu, J. S., Cao, Y., Li, S. C. (2007). Ouyang, H. Ecological risk assessment of regions along the roadside of the Qinghai-Tibet highway and railway based on an artificial neural network, *Human and Ecological Risk Assessment*, 2007, 13(4): 900-913.

CORDES (2014). Community Research and Development Information Service, European Commission FP7. Can be assessed at: http://cordis.europa.eu/fp7/home_en.html

Crews, S.C., Stevens, L.E. (2009). Spiders of Ash Meadows National Wildlife Refugge, Nevada. *The Southwestern Naturalist* 54:331-340.

Dubash N.K. (2002). Tubewell capitalism: groundwater development and agrarian change in Gujarat. 300 pp.

ISBN 019-565747-0, Can be accessed at: http://www.cabdirect.org/abstracts/20036792650.html;jsessionid=C60C6551CE86E05B708937AD A6F53E75

Echelle, A.A., E.W. Carson, A.F. Echelle, R.A. Van Den Bussche, T.E. Dowling, A. Meyer and R. M. Wood. (2005). Historical biogeography of the New-World pupfish genus *Cyprinodon* (Teleostei: Cyprinodontidae). Copeia 2005:320-339.

Eamus, D., Froend, R., Loomes, R., Hose, G., Murray, B. (2006a). A functional methodology for determining the groundwater regime needed to maintain the health of groundwater-dependent vegetation. Australian Journal of Botany 54, 97–114.

Eamus, D., Hatton, T., Cook, P.G., Colvin, C. (2006b). 'Ecohydrology: vegetation function, water and resource management.' (CSIRO: Melbourne).

Eamus, D., Froend, R. (2006). Groundwater-dependent ecosystems: the where, what and why of GDEs. *Australian Journal of Botany* 54, 91–96.

Ghanaian WRC 2011. Groundwater Management Strategy. http://doc.wrc-gh.org/pdf/wrc_4014e_20120227_1330329304_.pdf, Accessed January 22, 2014.

Gibb, H.A.R. (1969). "The Rise of Saladin, 1169–1189", A History of the Crusades: The First Hundred Years (2nd ed.), London: University of Wisconsin Press, pp. 563–589.

Grand Canyon Wildlands Council, Inc. (2002). An Assessment of 100 Seeps, Springs, and Natural Ponds on the Arizona Strip, Northern Arizona. Arizona Water Protection Fund, Arizona Department of Water Resources, Phoenix.

Hart, E.R. (1986). The barefoot trail: access to Zuni Heaven. Institute of the North American West, University of Utah, Salt Lake City, UT, 191 pp.

Hatton, T. and R. Evans. (1998). Dependence of ecosystems on groundwater and its significance to Australia. Occasional Paper No. 12/98. Canberra: Land and Water Resources Research and Development Corporation.

Haynes, C.V., Jr. (2008). Quaternay cauldron springs as paleoecological archives. Pp. 76-97 *in* Stevens, L.E. and V.J. Meretsky, editors. Aridland Springs in North America: Ecology and Conservation. University of Arizona Press, Tucson.

Heilweil, V.M., Sweetkind, D.S., Gerner, S.J (2014). Innovative Environmental Tracer Techniques for Evaluating Sources of Spring Discharge from a Carbonate Aquifer Bisected by a River. *Groundwater* Vol52, No.1, pp.71-83.

Hendrickson, D.A., J.C. Marks, A.B. Moline, E. Dinger, and A.E. Cohen. (2008). Combining ecological research and conservation: A case study in Cuatro Ciénegas, Coahuila, Mexico. pp. 127-157 *in* Stevens, L.E. and V.J. Meretsky, editors. Aridland Springs in North America: *Ecology and Conservation.* University of Arizona Press, Tucson.

Hershler, R. (1994). A review of the North American freshwater snail genus *Pyrgulopsis* (Hydrobiidae). *Smithsonian Contributions to Zoology* No. 554. Smithsonian Institution Press, Washington, DC.

Hershler, R. and T.J. Frest. (1996). A review of the North American freshwater snail genus *Fluminicola* (Hydrobiidae). *Smithsonian Contributions to Zoology*, 583: 1–41.

Hershler, R. and H-P Liu. 2008. Ancient vicariance and recent dispersal of springsnails (Hydrobiidae: Pyrgulopsis) in the Death Valley system, California-Nevada. Pp. 91-101 in Reheis, M.C., R. Hershler, and D.M. Miller. Late Cenozoic Drainage History of the Southwestern Great Basin and Lower Colorado River *Region: Geologic and Biotic Perspectives. Geological* Society of America Special Paper 439.

Hershler, R. and H-P Liu. (2012). Molecular phylogeny of the western North American pebblesnails, genus *Fluminicola* (Rissooidea: Lithoglyphidae), with description of a new species. *Journal of Molluscan Studies* 78:321-329.

Humphreys W. F. (2006). Aquifers: the ultimate groundwater-dependent ecosystems. *Australian Journal of Botany*, Vol. 54, Issue 2, p. 115-132.

IAH (2014). Website of the International Association of Hydrogeologists. Available at http://iah.org/groups/commissions-networks

Ingraham, N., K. Zukosky, and Kreamer, D.K. (2001). The Application of Stable Isotopes to Identify Problems in Large-Scale Water Transfer in Grand Canyon National Park. *Environmental Science and Technology*, 35-7 (1299-1302).

Johannessonn, K.H., Stetzenbach K. J. Hodge, V.F., Kreamer, D. K. Zhou, X. (1997). Delineation of Ground-Water Flow Systems in the Southern Great Basin Using Aqueous Rare Earth Element Distributions. *Ground Water* Vol. 35, No. 5, p. 807-819.

Khamala, C.P.M. (2005). Biodiversity and Sustainable Environmental Protection. In Science and Technology Capacity in the Framework of

Millennium Development Goals, Ed. BO Aduda, Proceedings of the National Workshop in Commemoration of the Scientific Revival Day, p. 63-72, December, 2005.

Kodrick-Brown, A. and J.H. Brown. (1993). Highly structured fish communities in desert Australian springs. *Ecology* 74:1847-1855.

Kodrick-Brown, A. and J.H. Brown. 2007. Native fishes, exotic mammals, and the conservation of desert springs. *Frontiers in Ecology and the Environment* 5:549-553.

Kreamer, D.K., Stetzenbach, K.J., Hodge, V.F., Johanneson, K. and I. Rabinowitz, 1996. Trace Element Geochemistry in Water from Selected Springs in Death Valley National Park, California. *Ground Water.* 34-1, p.95-103 (Jan-Feb. 1996).

Kresic, N. Stevanovic, Z. (2010). Groundwater Hydrology of Springs: *Engineering, Theory, Management and Sustainability*. Elsevier, Oxford.

Lazarus, P., 1997. Towards a Regulatory Framework for the Management of Groundwater in South Africa. Report to the Water Research Commission (WRC # 789/1/98) and the Department of Water Affairs and Forestry (Geo 2.2 – 389), Republic of South Africa, pp.65.

Ledbetter, J.D., L.E. Stevens, Hendrie, M.N. and A.E. Leonard, In press. Springs Ecosystems of Kaibab National Forest, Northern Arizona. Proceedings of the 12th Symposium on Research on the Colorado Plateau, Flagstaff.

Lewis, M.M., D. White, and T. Gotch, 2013. Spatial Survey and Remote Sensing of Artesian Springs of the Western Great Artesian Basin. National Water Commission, Canberra.

Liu, H-P, R. Hershler, and K. Clift, 2003. Mitochondrial DNA sequences reveal extensive cryptic diversity within a western American springsnail. Molecular Ecology 12: 2771-2782.

Liu, H-P. and R. Hershler, 2014. Microsatellite primers for a western North American springsnail (Pyrgulopsis robusta): evidence for polyploidy and cross-amplification in P. bruneauensis. *Journal of Molluscan Studies*, 80(1): 107-110. doi:10.1093/mollus/eyt050

Liu, H-S, R. Hershler, B. Lang, and J. Davies, 2013. Molecular evidence for cryptic species in a narrowly endemic western North American springsnail (*Pyrgulopsis gilae*). *Conservation Genetics* 14:917–923.

Madden, T., 2000. A Concise History of the Crusades. Rowman and Littlefield Publishers, Lanham, Maryland, USA. ISBN 978-0-8476-9430-3.

Makurira, H. and M. Mugumo, 2003. Water Sector Reforms In Zimbabwe: The Importance of Policy and Institutional Coordination On

Implementation, Food and Agriculture Organization of the United Nations, Chapter 14 in: *Proceedings of the African Regional Workshop on Watershed Management* p.167-174

McClain M.E., Boyer E.W., Dent C.L., Gergel S.E., Grimm N.B., Groffman P.M., Hart S.C., Harvey J.W., Johnston C.A., Mayorga E., McDowell W.H., and G. Pinay, 2003. Biogeochemical hot spots and hot moments at the interface of terrestrial and aquatic ecosystems. *Ecosystems* 6:301-312

Meinzer, OE. 1927. Large springs in the United States. Water Supply Paper 557, U.S. Geological Survey, Washington.

Meretsky, V.J., Stevens, L.E. 2000. Kanab ambersnail, an endangered succineid snail in southwestern USA. Tentacle 8:8-9.

Munch, Z. ; J Conrad, (2007). Remote sensing and GIS based determination of groundwater dependent ecosystems in the Western Cape, South Africa. *Hydrogeology Journal*, February 2007, Volume 15, Issue 1, pp 19-28.

Murphey, N.P., M.F. Breed, M.T. Guzik, S.J.B. Cooper, and A.D. Austin. 2012. Trapped in desert springs: phylogeography of Australian desert spring snails. Journal of Biogeography 39: 1573–1582.

Murray B.R., Zeppel M.J.B., Hose, G.C. and D. Eamus. Grounder-dependent ecosystems in Australia: It's more than just water for rivers. *Ecological Management and Restoration*, Vol. 4, No. 2, August 2003, pp.110-113.

Mutiga, Mutuku, J., 2005. Essentials and Targets of Water Resources Management in Kenya. In Science and Technology Capacity in the Framework of Millennium Development Goals, Ed. BO Aduda, Proceedings of the National Workshop in Commemoration of the Scientific Revival Day, p. 73-80, December, 2005.

Nature Conservation Council, 1999. Desktop Methodology to Identify Groundwater Dependent Ecosystems. Nature Conservation Council of New South Wales, Australia, Inc. PKK Environment and Infrastructure PTY Ltd, Concord West, NSW.

Nelson, N. 2008. Between the cracks: Water law and springs conservation in Arizona. Pp. 318-331 *in* Stevens, L.E. and V.J. Meretsky, editors. Aridland Springs of North America: *Ecology and Conservation.* University of Arizona Press, Tucson.

Odame-Ababio, K., 2003. Putting Integrated Water Resource Management Into Practice – Ghana's Experience, Food and Agriculture Organization of the United Nations, Chapter 13 in: *Proceedings of the African Regional Workshop on Watershed Management* p.157-166.

Odum, HT. 1957. Trophic structure and productivity of Silver Springs, Florida. *Ecological Monographs*, 27: 55-112.

O'Shea, T.J., and M.E. Ludlow. 1992. Florida manatee. Pp. 190-200 in Humphrey, S.R, editor. Rare and Endangered Biota of Florida. University Press of Florida, Gainesville.

Perez, K.E., W.F. Ponder, D.J. Colgan, S.A. Clark, and C. Lydeard. 2005. Molecular phylogeny and biogeography of spring-associated hydrobiid snails of the Great Artesian Basin, Australia. *Molecular Phylogenetics and Evolution* 34:545–556.

Perla, B.S. and L.E. Stevens. (2008). Biodiversity and productivity at an undisturbed spring in comparison with adjacent grazed riparian and upland habitats. Pp. 230-243 *in* Stevens, L.E. and V.J. Meretsky, editors. Aridland Springs in North America: Ecology and Conservation. University of Arizona Press, Tucson.

Phillips, D.A., Jr., C.S. Vanpool, and T.L. Vanpool. (2006). The horned serpent tradition in the North American Southwest. Pp. 17-30 in Religion in the Prehistoric Southwest. AltaMira Press, Lanham, MD.

Polhemus, D.A., Polhemus, J.T. (2002). Basins and ranges: The biogeography of aquatic true bugs (Insecta: Heteroptera) in the Great Basin. Pages 235-254 in Hershler, R., D.B. Madsen, and D.R. Currey, editors. Great Basin Aquatic Systems History. Smithsonian Contributions to the Earth Sciences No. 33, Washington, DC.

Ramsar (2014). Ramsar Website. Available online at: http://www.ramsar.org/cda/en/ramsar-about-about-ramsar/main/ramsar/1-36%5E7687_4000_0__

Ramsar (2008). Culture and Wetlands, a Ramsar Guidance Document, Gland, September (2008). Available online at: www.ramsar.org/pdf/.../cop10_culture_group_e.pdf

Rea, A.M. (2008). Historic and prehistoric ethnobiology of desert springs. Pp. 268-278 *in* Stevens, L.E. and V.J. Meretsky, editors. Aridland Springs in North America: Ecology and Conservation. University of Arizona Press, Tucson.

Remote Sensing Applications Center. (2012). Groundwater-Dependent Ecosystem Inventory Using Remote Sensing. Remote Sensing Applications Center RSAC-10011-RPT1, Salt Lake City.

Sada, D. W., Fleishman, E., and Murphy, D. (2006). Effects of environmental heterogeneity and disturbance on the native and non-native flora of desert springs. *Biological Invasions* 8:1091-1101.

Salman, S.M.A. (2009). The World Bank Policy for Projects on International Waterways, A Historical and Legal Analysis, The International Bank for

Reconstruction and Development/The World Bank, 1818 H Street NW, Washington D.C., USA,

Scofield, C.I. (Ed.) 1967. New Scofield Reference Bible, Authorized King James Version. 2 Chronicles32:1-4. Oxford University Press, Oxford, United Kingdom.

Sinclair Knight Mertz (2001). Australian Groundwater Dependent Ecosystems. Report by Sinclair Knight Mertz for Environment Australia, Armadale, Victoria, Australia.

Springer, A.E. and Ledbetter, J.D. (2012). Spring Mountains National Recreation Area Springs Inventory. Humboldt-Toiyabe National Forest 10-CR-11041705-052, Las Vegas, Nevada.

Springer, A.E. and L.E. Stevens, L.E. (2008). The sphere of discharge of springs. *Hydrogeology Journal* 17:83-93.

Springs Stewardship Institute. (2014). Springs Stewardship website. Available on-line at: www.springstewardship.org

Stevens, L.E., Ayers, T.J. 2002. The biodiversity and distribution of alien vascular plant and animals in the Grand Canyon region. Pp. 241-265 *in* Tellman, B. (Ed.) Invasive Exotic Species in the Sonoran Region. University of Arizona Press, Tucson.

Stevens, L.E. Meretsky, V.J. (Eds.) 2008. *Arid land Springs in North America: Ecology and Conservation*. University of Arizona Press, Tucson.

Stuebe, C., Richter, S. Griebler, C. (2009). First attempt towards an integrative concept for the ecological assessment of groundwater ecosystems. *Hydrogeology Journal* 17: 23-35.

Tennessee Valley Authority v. Hill (1978). 437 U.S. 153, 184 (1978). Available online at: https://supreme.justia.com/cases/federal/us /437/153 /case.html.

USFS (2010). Groundwater Dependent Ecosystems Inventory and Monitoring Protocols - Business Requirements Analysis, U.S. Forest Service January 30, 2010, Washington, DC. Available online at: http://www.fs.fed.us /geology/bus_require_analysis_v5_2[1].pdf

USFS (2012a). Groundwater Dependent Ecosystems: Level I Inventory Field Guide, Inventory Methods for Assessment and Planning. U.S. Forest Service Gen. Tech. Reort WO-86a March 2012, Washington, DC. Available online at: http://www.fs.fed.us/geology/GDE_Level_I_FG_ final_March2012_rev1_printing.pdf

USFS (2012b). Groundwater Dependent Ecosystems: Level II Inventory Field Guide, Inventory Methods for Project Design and Analysi. U.S. Forest Service Gen. Tech. Reort WO-86b March 2012, Washington, DC.

Available online at: http://www.fs.fed.us/geology/GDE_Level_II_FG_final_March2012_rev1_s.pdf

U.S. Forest Service. (2012c). Monitoring and Evaluation for Conserving Biological Resources of the Spring Mountains National Recreation Area: Final Program Report for 2010-2012. U.S. Department of Agriculture Forest Service, Rocky Mountain Research Station Grassland, Desert and Shrubland Program Forestry Sciences Laboratory, Albuquerque, NM.

USLegal (2014). Standing law and legal definition. Available online at: http://definitions.uslegal.com/s/standing/

Wethington, A.R., Lydeard, C. (2007). A molecular phylogeny of Physidae (Gastropoda: Basommatophora) based on mitochondrial DNA sequences. *Journal of Molluscan Studies* 73:241-257.

Whiting G.J., Chanton, J.P. (2001). Greenhouse carbon balance of wetlands: methane emission versus carbon sequestration. *Tellus* 53(B), 521-528.

Williams, C.D. (1984). The Decline of Ash Meadows, a Unique Desert Ecosystem. pp. 716-720 *in* Warner, R.E. and K.M. Hendrix, editors. California Riparian Systems: University of California Press, Berkeley. Available online at: http://ark.cdlib.org/ark:/13030/ft1c6003wp/

Witcombe (1998). Water and the Sacred. Written and produced by C.L.C.E. Witcombe, Department of Art History, Sweet Briar College, Virginia 24595 USA. Available online at: http://witcombe.sbc.edu/sacredplaces/water.html

Reviewed by Joseph Gurrieri, Chairman of IAH Network on Groundwater and Ecosystems, National Groundwater Program Leader, USDA Forest Service, 740 Simms St., Golden CO 80401 and Mike Wireman, co-chairman of IAH Network on Groundwater and Ecosystems.

In: Groundwater
Editor: Segun Michael Adelana
ISBN: 978-1-63321-759-1

Chapter 10

BIOREMEDIATION OF CHLOROETHYLENES: AN EVALUATION OF GROUNDWATER RECIRCULATION SYSTEMS

H. M. T. de Vries* and A. A. M. Nipshagen
Bioclear, Groningen, The Netherlands

ABSTRACT

From more than ten years of experience with a groundwater recirculation system called the TCE-concept, it can be concluded that bioremediation is a valuable technique to remediate groundwater contaminated with chlorinated aliphatic hydrocarbons (CAHs). The most prominent process during bioremediation is reductive dechlorination. Reductive dechlorination is a process resulting in the detoxification of CAHs. This process can be catalyzed by a bacterial consortia containing *Dehalococcoides* spp (DHC).

With the TCE-concept, the reductive dechlorination of CAHs is stimulated by mixing DHC, lactate, acetate and nutrients with CAH contaminated groundwater. The success of the TCE-concept is highly dependent on the distribution of the amendments within the targeted treatment zone. For the distribution a recirculation system is installed in a permeable soil layer. Unsuccessful remediation is the result of uneven distribution of the amendments, caused by for instance, improper well placing and unexpected impermeable clay or peat layers.

* Corresponding author: deVries@bioclear.nl.

The following are characteristics of the TCE-concept:

- The TCE-concept is applicable in rural as well as urban or densely built areas.
- Soil volumes of 12,000 to 400,000 m^3 have been treated.
- CAH concentrations up to 10% of the saturation concentration can be treated.
- CAH concentration of <10 μg/l and a CAH load reduction of 99.9% can be achieved.
- Biological degradation of CAHs takes one to three years.
- The costs of € 3.90 to € 5.50 per cubic meter are low compared to other techniques.
- Bioremediation has a low environmental impact compared to other techniques.

Insight in the chemical and biological processes in the soil is crucial for successful bioremediation. To gain insight, a database with data from more than 300 sampling wells from nine TCE-concept sites has been analysed. From the data analysis it can be concluded that during the TCE-concept, the TOC concentration will be elevated for approximately 300 days. All sulphate is reduced within the first 200 days, creating strictly reduced conditions, favourable for CAH degradation. The CAHs are degraded as soon as elevated TOC concentrations are present, but the degradation process gains momentum as soon as all the sulphate is removed. As long as TOC is present, methane is produced. Enough organic carbon has to be dosed during the TCE-concept to reduce all the sulphate and CAHs. Some extra organic carbon has to be dosed in order to compensate for methanogenic activity.

INTRODUCTION

The solvents tetrachloroethylene (PCE) and trichloroethylene (TCE) have extensively been used in the past by (metal) industries and drycleaners. Due to improper storage or disposal of the solvents, significant amounts got spilled and migrated to the groundwater. PCE, TCE and their subsequent degradation products dichloroethylene (DCE) and vinyl chloride (VC) migrate easily in groundwater and are relatively recalcitrant to biodegradation, resulting in large groundwater areas contaminated with chlorinated aliphatic hydrocarbons (CAHs). Due to their toxic and carcinogenic properties, CAHs pose a serious risk to human health. To reduce these risks, and restore the possible use of groundwater as a risk-free resource, groundwater remediation is necessary.

The optimal trade-off between costs and remediation results for conventional remediation techniques, like pump and treat or excavation, are tough to find. Excavation can remove the entire contamination, but at high costs. Pump and treat can be cheaper on the short run, but in many cases, it leaves unwanted residual contamination, resulting in expensive elongation of the remediation time (Semprini, 1995). During the past decade a new efficient and effective remediation technique has been developed: bioremediation. Bioremediation uses microorganisms to remediate soil and groundwater contaminants.

In the past decade the number of bioremediation projects has increased steadily in the USA (Lyon and Vogel, 2013) and the Netherlands. These projects have contributed to a large body of knowledge. In this chapter, more than ten years of experience with the bioremediation technique called the Total Concept Evenblij (TCE-concept) is described. The TCE-concept combines a continuous-fed bioreactor with a groundwater recirculation system. This concept has been developed by Bioclear in cooperation with the local government of Drenthe and the Dutch Centre of Soil Quality Management and Knowledge transfer (SKB) (Henssen, 1999).

The knowledge of more than ten years of bioremediation with the TCE-concept is discussed in this chapter, giving special attention to:

- reductive dechlorination;
- design of the TCE-concept;
- remediation results and costs;
- chemical and biological processes during bioremediation;
- environmental impact.

REDUCTIVE DECHLORINATION

A prominent biodegradation route of CAHs is reductive dechlorination. Reductive dechlorination is a process in which a chlorine atom is replaced by a hydrogen atom. This process can be catalyzed by microorganisms. The reductive dechlorination of PCE is depicted in Figure 1.

The process of reductive dechlorination has been intensively investigated (Smidt and de Vos, 2004). In the reductive dechlorination process, the chlorine atoms are replaced step by step, reducing PCE via TCE, DCE and VC to the harmless end products ethylene and ethane. The CAHs act as electron acceptors. Hydrogen, which can be produced from organic carbon, acts as an electron donor.

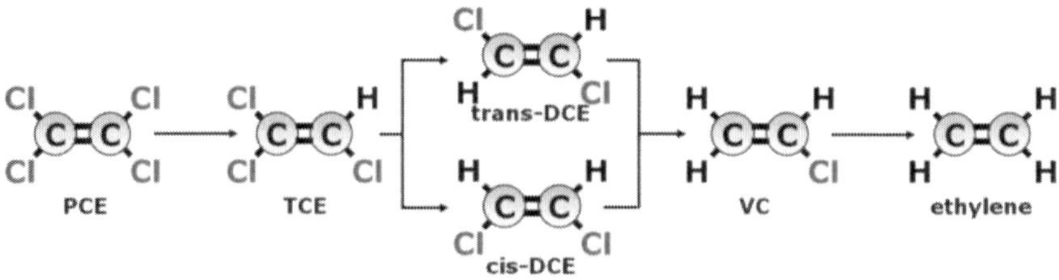

Figure 1. Reductive dechlorination of PCE.

Reductive dechlorination of PCE and TCE to DCE occurs under iron reducing to methanogenic redox conditions. The intermediates DCE and VC can be degraded to ethylene or ethane only under the following conditions:

1. Strictly reduced (methanogenic) conditions.
2. Availability of sufficient organic carbon.
3. Presence of dechlorinating microorganisms, *Dehalococcoides* spp. (DHC).

If the reductive dechlorination process does not occur at the contaminated site, or it stalls at DCE or VC, then addition of amendments can stimulate the process. To create methanogenic conditions, and to reduce oxygen, nitrate, manganese, iron and sulphate, organic carbon needs to be added. Examples of organic carbon used in bioremediation are whey, molasses, ethanol and lactate. When enough organic carbon is added, methanogenic conditions will be created and the remainder of the organic carbon will act as an electron donor for the reductive dechlorination process.

The key in achieving complete reductive dechlorination is a bacterial consortia containing the species *Dehalococcoides* spp. (DHC). DHC is the only known bacterial species that is able to completely degrade PCE to ethylene via reductive dechlorination (Freedman and Gossett, 1989). By introducing and/or stimulating the growth of DHC, complete degradation of PCE can be achieved (Stroo, 2010). DHC can be introduced in the groundwater by infiltrating DHC that is either cultured in a lab, or that is cultured on site in a bioreactor. The presence of DHC can be demonstrated with molecular detection techniques like qPCR.

TCE-CONCEPT DESIGN

With the TCE-concept, the reductive dechlorination of CAHs is stimulated by mixing DHC and organic carbon with CAH contaminated groundwater. A schematic overview of the TCE-concept is presented in Figure 2.

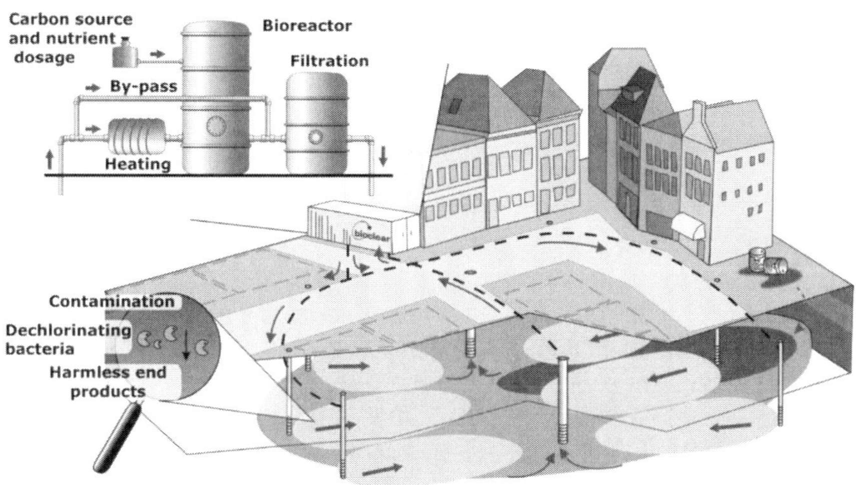

Figure 2. Schematic overview of the TCE-concept.

The DHC is cultured in an anaerobic bioreactor on site. The influent of the bioreactor is contaminated groundwater from the site. The effluent of the bioreactor contains more than $1*10^4$ cells DHC/ml.. The bioreactor effluent is filtered for suspended solids and infiltrated back into the groundwater system, together with an organic carbon mixture and nutrients (nitrogen and phosphorous).

The organic carbon mixture contains lactate and acetate. Cheaper organic carbon substrates like whey or molasses cannot be used because they contain suspended solids, making long-term infiltration impossible due to well clogging problems. To further reduce the risk of well clogging, the lactate/acetate mixture is dosed only once a day during a short period of time. Furthermore the low pH of the mixture results in unfavourable growth condition in the infiltration wells.

The phase in which active groundwater extraction and infiltration takes place, is called the active phase of the bioremediation. When the amendments have been distributed to the targeted treatment zone, the bioreactor is removed, and the passive phase starts. In the passive phase the CAHs are degraded to ethylene and ethane.

Remediation Results of the TCE-Concept

Since the first full scale application of the TCE-concept in 1998, the concept has been applied at more than ten sites. The site descriptions are summarised in Table 1. In the following paragraphs, the sites' characteristics and remediation results are discussed.

Treated Soil Types

The success of the TCE-concept is highly dependent on the distribution of the amendments within the targeted treatment zone. The soil has to be permeable in order to evenly distribute the amendments with the recirculation system. Based on experience, a minimum hydraulic conductivity (K_d) of 1 m/d is required. The soil types treated at the sites summarised in Table 1 have a relatively high hydraulic conductivity (sand or limestone).

At several sites the treated soil layer is bound by poorly permeable clay or peat layers (not shown in the table). If clay or peat layers are present in the treated area, they increase the risk of incomplete distribution of amendments. Incomplete distribution needs to be prevented because this can result in untreated pockets of high CAH concentration, recontaminating the site after the remediation operation has ceased.

The risk of peat or clay layers is demonstrated at the site "Wilgstraat – The Hague (NL)". At this site, unexpected clay layers prevented even distribution of amendments. This resulted in an untreated area, even after prolongation of the active phase. The final concentrations were therefore higher than expected.

Treated Soil Volume, Remediation Duration and Well Configuration

Table 1 shows that the treated soil volumes are relatively large and vary between 12,000 and 400,000 m^3. To distribute the amendments within the targeted treatment zone, a recirculation system is used.

Table 1. Summary of sites where the TCE-concept has been applied

Site location	Treated soil type (m-gl)	Treated soil volume (m^3)	Duration active phase (A), passive phase (P) (months)	Highest Initial CAH concen-tration (µg/l)	Highest end CAH concen-tration (µg/l)	Total % CAHs removed	Costs (€)
Hoogeveen (NL)	Sand (9 – 40)	400,000	A: 18 P: 22	50.000	< 10	99.9 %	1,500,000
Almelo (NL)	Sand (3 - 11)	12,000	A: 5 P: 12	15.000	< 1	99.9 %	80,000
Twello (NL)	Sand (2 – 13)	50,000	A: 10 P: 24	7.500	2.300	85 %	300,000
Kongevej (DM)	Limestone (8 – 16)	20,000	A+P: 4	20	< 0,1	99 %	100,000
Drie bergen (NL)	Sand (3 – 6)	40,000	A: 6 P: 22	2.500	< 10	99 %	200,000
Gouds-bloemlaan – The Hague (NL)	Sand (2 – 10)	80,000	A: 9 P: 21	1.000	< 10	97 %	440,000
Wilgstraat – The Hague (NL)	Sand (2 – 18), loam layer present (8 – 9)	285,000	A: 24 P: 24	1.500	180	85 %	1,200,000
Boreelstraat – The Hague (NL)	Sand (6 – 16)	225,000	A: 27 P: 24	10.000	< 2,5	99,9 %	885,000
Newton-straat – The Hague (NL)	Sand (4 – 17), clay layer present (9 – 10)	70,000	A: 6 P: 22	1.000	< 10	99 %	350,000

Note: Table 1 does not contain two sites where the TCE-concept is still in progress: Vorden (NL) and P. Buijsstraat, The Hague (NL).

A recirculation system is preferred over batch amendment approaches because it makes fast and accurate distribution possible.

On average, the soil pore volume of the treated area needs to be flushed 1.5 times. The well configuration of the recirculation system depends on local site characteristics. With modelling software, like Visual Modflow and MT3D, the optimal well configuration which results in a fast and even distribution of adequate organic carbon concentrations, can be determined. Figure 3 shows a circular well configuration.

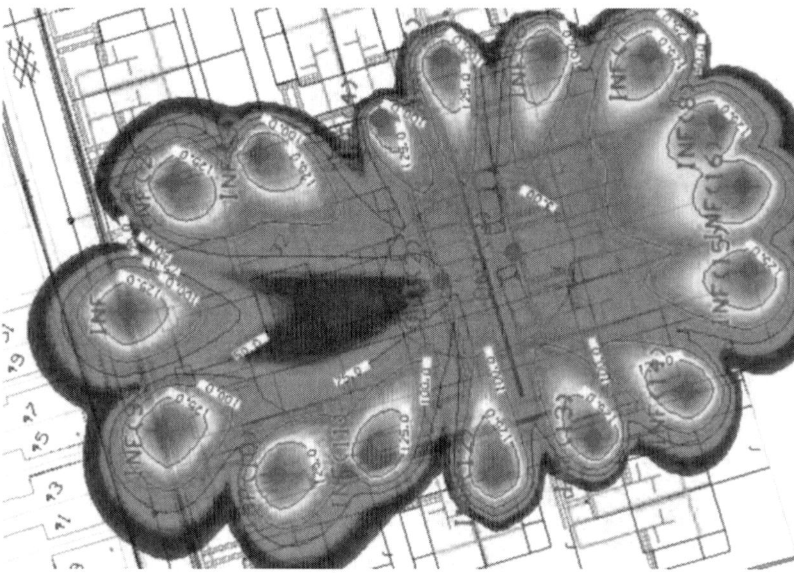

Figure 3. Circular well configuration (Visual Modflow picture of organic carbon concentrations at the end of an infiltration period).

Figure 4 shows a cascaded well configuration. In the cascaded well configuration, extraction wells are reused in the next phase as infiltration wells, making efficient and effective distribution of the amendments over a larger area possible without having to increase the bioreactor's capacity.

With the TCE-concept the remediation is performed *in situ*, making large excavations or local demolition of buildings unnecessary. The distance between the infiltration and extraction wells can be up to 80 meters, making the TCE-concept applicable in urban or densely built areas with limited space for remediation operations. Larger distances between the infiltration and extractions wells might be possible if a carbon source is used that does not degrade as fast as the acetate/lactate mixture. Fam (2012) uses an ethanol/vegetable oil mixture to combine fast degrading and slow degrading carbon sources. However no experience is available with this mixture in combination with the TCE-concept, and it is doubted whether oil can be distributed as easily as lactate, acetate or ethanol.

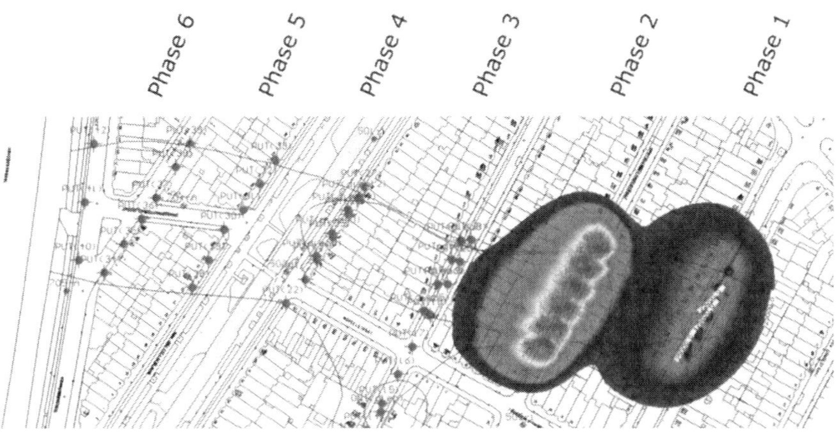

Figure 4. Cascaded well configuration (Visual Modflow picture of organic carbon concentrations at the end of phase 2).

The site "Twello (NL)" shows the importance of a thorough understanding of the contamination situation, and an adequate well configuration. After the active phase, it appeared that due to an inadequate well configuration, part of the contaminated area had not been treated. Nevertheless, in the untreated zone, a positive influence on the CAH concentration from the treated zone was measured, but this was not enough to reach the aimed remediation goal.

The active phase of the sites summarized in Table 1 varied between four to twenty seven months, depending on the size of the site. After the active phase, the bioreactor is removed from the site and the passive phase starts. Generally the passive phase takes one to three years.

Treated CAH Concentrations

Table 1 shows that the treated CAH concentrations vary between 20 and 50.000 μg/l. Contamination concentrations up to 10% of the saturation concentration can be treated with the TCE-concept. Literature shows that reductive dechlorination is also possible at CAH concentrations near the saturation concentration (Adamson et al., 2003). But if CAH concentrations are present in concentrations above 10% of the saturation concentration it is likely that the CAHs are present as dense, non-aqueous phase liquids (DNAPL). DNAPLs behave differently than dissolved CAHs. DNAPLs are heavier than water and migrate downwards, creating 'pools' of DNAPL on poorly permeable soil layers, leaving a trail of residual DNAPL on its way down. DNAPLs cannot be directly degraded by microorganisms because of its toxicity. First the DNAPL has to dissolve.

A zone of high biological activity around a DNAPL can accelerate the dissolution process. But if a significant pool of DNAPL is present, the dissolution process will be too slow to be remediated within the time frame of the TCE-concept. If DNAPLs are present at the site, it is advised to use the TCE-concept for the plume and local slow-release electron donor injections at the DNAPL zone. The injections can be performed with a direct push technology. A cheap and reliable slow-release electron donor is for instance protamylasse, a by-product of the potato starch industry. One or more injection rounds ensure that the DNAPL is 'encapsulated' by a biologically active zone, depleting the DNAPL pool over time.

Remediation Results

Table 1 shows that a successful TCE-concept remediation results in CAH concentration of 10 μg/l or lower. The total amount of CAH is reduced by 97 to 99.9%.

As mentioned earlier, the remediation at the sites "Wilgstraat – The Hague (NL)" and "Twello (NL)" did not reach the aimed remediation goal due to unexpected clay layers and improper well configurations. The experiences gained on these sites stress the importance of a proper design of the TCE-concept to ensure even distribution of the amendments throughout the targeted treatment zone.

Remediation Costs

The costs of the TCE-concepts applied at the sites summarised in Table 1 vary between € 80,000 - and € 1,500.000 -. The first TCE-concepts were relatively expensive due to their innovative nature. In the past decade, the TCE-concept has been improved becoming more effective and efficient. The most recent TCE-concepts require € 3.90 to € 5.50 per cubic meter contaminated soil, depending on their size..

Compared to extractive techniques, like pump and treat or air sparging, the TCE-concept is more cost effective and more sustainable because of the absence of long term above ground installations necessary to remediate the extracted groundwater / air.

Chemical and Biological Processes during Bioremediation

A proper design is a crucial first step for the success of the TCE-concept. Once the active phase is started, groundwater monitoring is needed to check whether the design works out as planned. For the groundwater monitoring, sampling wells are placed and sampled to monitor the following processes:

- the distribution of organic carbon;
- the creation of strictly reduced (methanogenic) conditions;
- the presence of dechlorinating microorganisms;
- the degradation of CAHs.

In practise, mostly a limited number of sampling wells is available to monitor a site. Interpreting data from only a few sampling wells and interpolating this to the whole site is difficult. Insight in the chemical and biological processes that occur during bioremediation is crucial to be able to draw the right conclusions.

To give insight into the chemical and biological processes, data from the nine TCE-concepts summarized in Table 1 have been analysed. The initial database consisted of 318 sampling wells. Not all wells were located in the treated area, therefore a sampling well selection was made. The criterion for the sampling well selection was: at least once the organic carbon concentration should be higher than 100 mg/l. This criterion ensures that the sampling well is located in the treated area and that the chemical and biological processes which are typical for bioremediation will occur. The selection resulted in 74 sampling wells with a total of 392 data points. Most data points contain results for the following parameters: redox potential, pH, conductivity, total organic carbon (TOC), oxygen, nitrate, sulphate, methane, DHC, CAHs and the CAH degradation products ethylene and ethane.

To be able to compare the data series of the 74 sampling wells, the data series have been aligned in time. The date at which the highest TOC concentration of the data series is measured is renamed as t = 0. So for each well the actual bioremediation started before t = 0, at t = 0 the highest TOC concentration was measured and from t = 0 onwards the chemical and biological processes occur.

In the following paragraphs the interpretation of the results is described. The data are depicted in graphs which were made with the statistical programme 'R', graphing package 'ggplot'. The line in the graph is the smooth average value, the gray area above and below the average value is the confidence interval (0.95).

Distribution and Degradation of Organic Carbon

Organic carbon is added to achieve TOC concentrations of 200 to 400 mg/l in the groundwater (depending on the soil chemistry). The distribution and degradation of organic carbon over time is depicted in Figure 5.

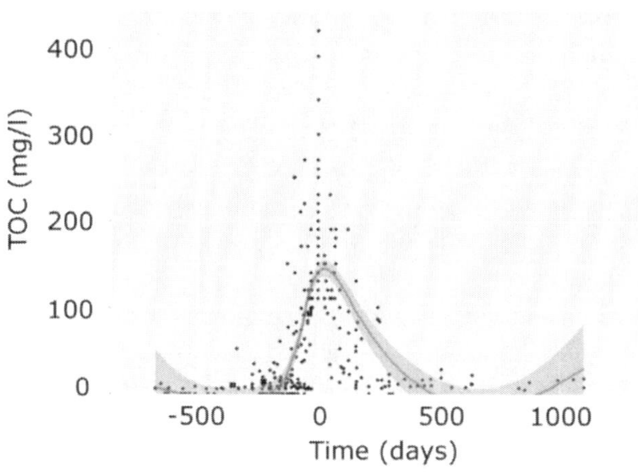

Figure 5. Distribution and degradation of lactate and acetate mixture (t=0 is the time at which the highest TOC concentration in the sampling wells are measured).

Figure 5 shows that the initial TOC concentrations vary between 0 and 10 mg/l. At t= -100 days, the first increase in TOC concentrations is visible, indicating the start of the active phase of the TCE-concept. Because of the data alignment the highest TOC concentrations are reached at t = 0. At this measuring point not all groundwater samples contain (the aimed concentration of) 200 to 400 mg/l TOC. This is because the heterogeneous soil profile results in uneven distribution of organic carbon at the site Sampling wells in which TOC concentration of 100 mg/l are not reached need to be monitored carefully to make sure that these low TOC concentrations are enough for complete CAH degradation. If a soil pocket with low TOC concentration is surrounded by a large body of groundwater with high TOC concentration, complete VOCl degradation is expected over time due to diffusion of TOC.

After the TOC peak is achieved, degradation of TOC takes place. Figure 5 shows that the TOC concentrations decrease after t = 0 to approximately 50 mg/l at t = 50 days. At t = 200 days almost all TOC is degraded, and a steady state between 0 and 20 mg/l TOC is reached. In total, elevated TOC concentration have been measured from t = - 100 to t = 200 (300 days).

Creation of Strictly Reduced (Methanogenic) Conditions

By adding organic carbon as an electron donor, electron acceptors like oxygen, nitrate, manganese, iron and sulphate are reduced. Once these compounds are reduced, favourable conditions are created for reductive dechlorination and methanogenic activity. The methanogenic activity results in the formation of methane.

To follow the redox conditions in a sampling well, it is not necessary to measure the oxygen, nitrate, manganese, iron, sulphate and methane concentration. In practise, only sulphate and methane are sufficient to get a good idea of the redox conditions, and whether these are favourable for reductive dechlorination. If sulphate is completely reduced to sulphide, it is assumed that the other electron acceptors like oxygen, nitrate, manganese and iron have also been reduced. When highly reduced conditions are reached, methane is formed. Decreasing sulphate concentrations (< 20 mg/l) and increasing methane concentrations (> 1 mg/l) are a clear sign of highly reduced conditions. The sulphate and methane concentrations during the TCE-concept are shown in Figure 6.

Figure 6 shows that the initial sulphate concentrations vary between 0 and 300 mg/l, with an average around 110 mg/l. As soon as the TOC concentration increases (t = -100 days), the sulphate concentration decreases. A sharp decrease is observed at t = 0. Between t = -100 and t = 100 (200 days) complete reduction of sulphate into sulfide is observed. Once the TOC concentration decreases, an increase in sulphate concentration is observed in a few wells. These rebound concentrations are significantly lower that the initial sulphate concentrations.

Figure 6 shows that the initial methane concentration is low, typically below 1 mg/l. The methane concentration start to increase slowly as soon as the TOC concentration increases. At the point when almost all sulphate is reduced, the methane concentration increases rapidly and reaches a concentration between 20 and 40 mg/l.

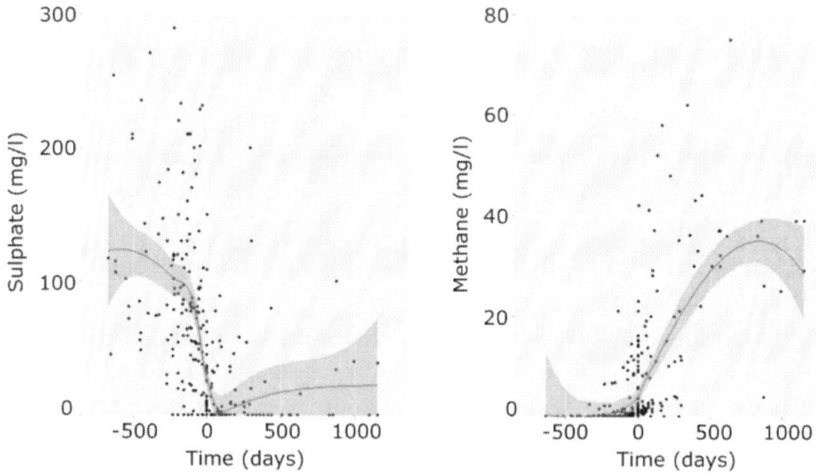

Figure 6. Sulphate and methane concentration during bioremediation (t=0 is the time at which the highest TOC concentration in the sampling wells are measured).

Favourable redox conditions for reductive dechlorination are characterized by a sulphate concentration lower than 20 mg/l and a methane concentration higher that 1 mg/l. Figure 6 shows that these conditions are achieved at t = 0, 100 days after the first increase in TOC is observed.

Presence and Growth of Dechlorinating Microorganisms

The presence of DHC is demonstrated by qPCR techniques. The aim of the TCE-concept is to infiltrate groundwater with a DHC concentration of $2*10^3$ cells DHC/ml. Relatively few DHC analyses have been performed on the sampling wells. From the limited data it can be concluded that at most sites, the initial DHC concentrations are lower than the detection limit. As soon as organic carbon and microorganisms are added, an increase in DHC concentrations is measured. DHC is transported through the soil as fast as the carbon source. Once the DHC is infiltrated together with de carbon source and nutrients, DHC will proliferate and reach concentration up to $1*10^6$ cells DHC/ml. It is not known how long the DHC will remain present in the groundwater.

Degradation of CAHs

The result of increased organic carbon concentrations, strictly reduced (methanogenic) conditions and the presence of dechlorinating microorganisms, is the complete degradation of CAH to the harmless compounds ethylene and ethane.

An aspect which provides information on the reductive dechlorination process of CAHs is the dechlorination grade. The dechlorination grade is expressed as a percentage. A percentage of 100% means that all CAHs present are the unreduced PCE. When reductive dechlorination takes place, lower chlorinated ethylenes like TCE, DCE and VC are formed. As the amount of the most heavily chlorinated ethylenes decrease, the dechlorination grade increases. When all the CAHs are degraded, and ethylene and ethane are formed, a dechlorination grade of 100% is achieved. The CAH concentrations and dechlorination grade during the TCE-concept are shown in Figure 7.

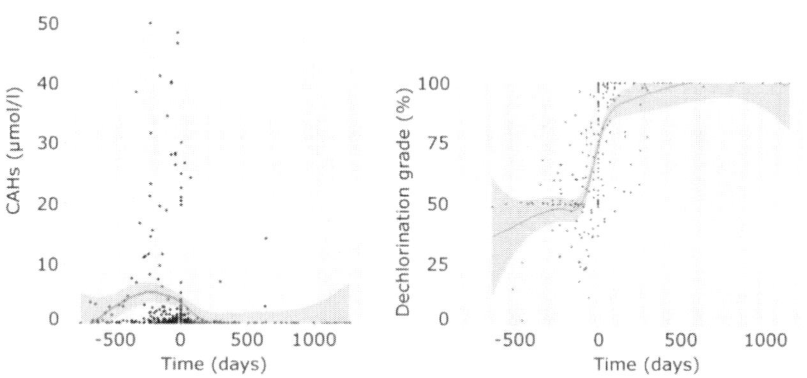

Figure 7. CAH concentration (total of PCE, TCE, DCE and VC) and dechlorination grade during bioremediation (t=0 is the time at which the highest TOC concentration in the sampling wells are measured).

Figure 7 shows that the initial CAH concentrations vary between 0 and 50 µmol/l. The graph suggests that at the start of the TCE-concept, the CAH concentration increases. Experience confirms changes in CAH concentrations directly at the start. But these changes are not necessarily only increasing concentrations, they sometimes decrease as well. Another explanation for the CAH peak at t = -250 days is the sampling well selection. It is possible that only one or two sites contain CAH data at t = - 500 days. These sites may be contaminated with CAH concentrations that are lower than in the other sites.

As soon as the TOC concentration increases (t = -100 days), the CAH concentration starts to decrease. From the raw data it is clearly visible that the dechlorination process starts with PCE, followed subsequently by TCE, DCE, VC, ethylene and ethane. The CAH degradation rate is fastest at t = 0 when the organic carbon concentrations peak. At t = 200 days, most CAHs have been degraded. Typical remediation results are CAH concentrations less than 10 μg/l, also see the next chapter for the remediation results. Due to an heterogeneous soil profile, it is possible that CAH concentration remain relatively high in local soil pockets. From experience, it appears that slow CAH degradation continues for up to three years, removing the last CAHs in the local soil pockets.

The dechlorination grade graph supports the above mentioned findings. Between t = - 100 and t = 100 days, the period of the highest TOC concentrations, the dechlorination grade increases from 40% to 90%. After t = 100 days, the dechlorination grade increases slowly from 90% to eventually 100%.

Acidification

When organic carbon is added, biological and chemical processes can result in acid formation. A favorable pH for DHC is between 6 and 8. Fam (2012) reports several sites in America that needed pH adjustments with potassium hydroxide or sodium hydroxide when treated with ethanol/whey mixtures.

From the results of the TCE-concept (not shown in this chapter), it is clear that the groundwater pH is not influenced to a degree that pH adjustments are needed.

Overview of Processes during Bioremediation

The graphs depicted in Figure 5, Figure 6 and Figure 7 have been combined into one graph, see Figure 8. This graph gives an impression of different compound concentrations over time.

Figure 8 shows that as soon as the TOC concentration increase, several processes start simultaneously: sulphate reduction, methane formation and CAH degradation. The process that 'finishes' first is the sulphate reduction. The second process to 'finish' is the CAH degradation. The CAH degradation starts as soon as organic carbon is present, but gains momentum as soon as the sulphate concentrations are low. The methanogenic activity also start as soon as organic carbon is present, and gains momentum as soon as the sulphate concentrations are low. The difference with the CAH degradation process is that the methanogenic activity lasts longer, continuing until all the organic carbon is consumed.

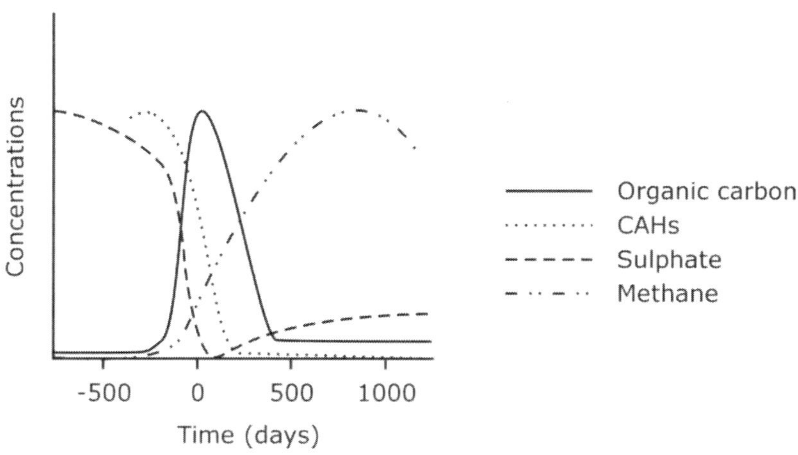

Figure 8. Impression of sulphate, CAH, TOC and methane concentrations during the TCE-concept remediation (t=0 is the time at which the highest TOC concentration in the sampling wells are measured).

From the results we can learn that enough organic carbon has to be added to reduce all the sulphate and CAHs. Some extra organic carbon has to be added in order to compensate for methanogenic activity. Overdosing organic carbon will result in more methane formation. Whether overdosing will help the CAH degradation is unclear.

SUSTAINABILITY

Nature has served as a source of inspiration for the design of the TCE-concept. In nature, PCE and TCE concentrations decrease due to physical, chemical and biological processes, also referred to as natural attenuation. Generally, natural attenuation is a slow process. The biological process of reductive dechlorination can however be accelerated using the TCE-concept by adding carbon source and DHC. After the amendments have been evenly distributed, the 'real' work is done by nature. The remediation effort and environmental impact of bioremediation is therefore low compared to other remediation techniques.

A life cycle assessment confirms the above mentioned by showing that the impact of bioremediation on the environment is low compared to thermal extraction and excavation (Lemming, 2010). Noteworthy is that bioremediation has even less impact on the environment than natural attenuation, because the potential future ecotoxological and human risks are considerably lower with bioremediation than with natural attenuation.

CONCLUSION

From more than ten years of experience it can be concluded that bioremediation with the TCE-concept is a valuable technique to remediate groundwater contaminated with CAHs. The TCE-concept stimulates reductive dechlorination of CAHs by mixing DHC, lactate, acetate and nutrients with CAH contaminated groundwater. Since the first full scale application of the TCE-concept the concept has been applied at more than ten sites.

The success of the TCE-concept is highly dependent on the distribution of the amendments to the targeted treatment zone. For the distribution of amendments a recirculation system is used. The soil has to be permeable enough in order to evenly distribute the amendments. The soil volumes treated with the TCE-concept are relatively large and vary between 12.000 and 400.000 m^3. The remediation is performed *in situ*, and the distance between the infiltration en extraction wells can be up to 80 meters, making the TCE-concept applicable in urban or densely built areas with limited space for remediation operations. The groundwater recirculation phase lasts for four to twenty seven months, depending on the size of the site.

After the recirculation phase, the bioreactor that cultures the DHC is removed from the site, and the bacteria degrade all the CAHs in one to three years. Contamination concentrations up to 10% of the saturation concentration can be treated with the TCE-concept. A DNAPL has to be treated with a separate DNAPL remediation technique. A successful TCE-concept remediation results in CAH concentration of 10 μg/l or lower and a CAH load reduction of 97 to 99.9%. Unsuccessful remediation is the result of uneven distribution of the amendments, for instance caused by, improper well placing and unexpected impermeable clay or peat layers. The remediation results and relatively low costs of € 3.90 to € 5.50 per cubic meter strike a good balance between remediation results and costs, compared to techniques like excavation, pump and treat or air sparging. Extra advantage of bioremediation is the low environmental impact.

Insight in the chemical and biological processes that occur during bioremediation is crucial for a successful implementation of bioremediation projects in general. To gain insight, a database of 318 sampling wells from nine TCE-concept sites has been analysed.

For complete CAH degradation it is important that strictly reduced (methanogenic) conditions, enough organic carbon and dechlorinating microorganisms are present. By applying the TCE-concept, the TOC concentration will increase for approximately 300 days. During the first 200 days, all sulphate present in the groundwater is reduced to sulphide, creating strictly reduced conditions. The CAHs are degraded as soon as elevated TOC concentrations are present, but the degradation process gains momentum as soon as all the sulphate is removed. As long as TOC is present, methane is produced. During the period with elevated TOC concentrations, most of the CAHs are degraded. From experience it is known that slow CAH degradation continues for up to three years, removing the last CAHs. From these results we can learn, amongst others, that enough organic carbon has to be added to reduce all the sulphate and CAHs. Some extra organic carbon has to be added in order to compensate for methanogenic activity.

ACKNOWLEDGMENTS

The authors would like to thank the following persons for their contribution to this chapter: D. de Groot, I. J. T. Dinkla, S. H. Lieten, S. Keuning, M. J. C. Henssen, H. F. Groenewold, A. S. Roosma and J. B. M. van Bemmel.

References

Adamson, D.T., McDade, J.M., Hughes, J.B. (2003). Inoculation of a DNAPL source zone to initiate reductive dechlorination of PCE. *Environmental Science and Technology*. 37, 2525-2533.

Bemmel, M. van, Van Der Werf, A., Henssen, M., Van Ras, N. (2007). Experiences with Bioaugmentation in Full-Scale Applications. In *International in situ and on site bioremediation symposium*. Columbus, Battelle Press. 1096-1103.

Fam, A. F., Falatko, D. M., Higgins, J. E., Mountain, S., Pirelli, A. J., Gaudette, M. (2012). A universal design approach for in situ bioremediation developed from multiple project sites. *Remediation*. 22(4), 49-74.

Freedman, D.L., Gossett, T.M. (1989). Biological reductive dechlorination of tetrachloroethylene and trichloroethylene of ethylene under methanogenic conditions. *Applied Environmental Microbiology*, 55, 2144-2151.

Henssen, M. J. C., Keuning, S., Werf, A. W. van der, Waarde, J. J. van der, Hubach, C., Blokzijl, R., Mourik, J., Haasnoot, C., Bosgoed, H., Meijerink, E. (1999). Full-scale bioremediation of chlorinated ethylenes: combining an anaerobic bioreactor with in situ stimulation. In Leeson, A., Alleman, B. C., eds. (1999) *Engineered approaches for in situ bioremediation of chlorinated solvent contamination*. Battelle Press, Columbus, Ohio, USA, 225-229.

Lemming, G., Hauschild M.Z., Chambon J., Binning P.J., Bulle C., Margni M., Bjerg P.L. (2010). Environmental Impacts of Remediation of a Trichloroethene-Contaminated Site: Life Cycle Assessment of Remediation Alternatives. *Environmental science & technology*. 44(23), 9163–9169.

Lyon, D.Y., Vogel, T.M. (2013). Bioaugmentation for groundwater remediation: an overview. In Stroo, H.F., Leeson, A., Ward, C.H., eds. (2013) *Bioaugmentation for groundwater remediation*. Springer, New York, NY, USA, 1-17.

Semprini, L. (1995). In situ bioremediation of chlorinated solvents. *Environmental Health Perspective,* 103(5), 101–105.

Smidt, H., de Vos, W. M. (2004). Anaerobic microbial dehalogenation. *Annual Review of Microbiology*. 58, 43-73.

Stroo, H.F., Major, D.W., Gossett, J.M. (2010). Bioaugmentation for anaerobic bioremediation of chlorinated solvents. In Stroo H.F., Ward

C.H., eds. *In Situ Remediation of chlorinated Solvent Plumes*. Springer, New York, NY, USA, 425-454.

Reviewed by S. Keuning (Bioclear, Groningen, The Netherlands) and S. Michael Adelana (Department of Environment & Primary Industries, Victoria, Australia)

EDITOR'S CONTACT INFORMATION

Dr. Segun Michael Adelana
Research Scientist – Hydrogeology
Victorian Department of Environment and Primary Industries
Agriculture Research Division, Australia
Email: Michael.Adelana@depi.vic.gov.au

INDEX

C

D

E

F

G

H

I

J

K

L

O

P

Q

R

S

T

U

V

W

Y

Z